The Expert System for Thermodynamics

A Visual Tour

Subrata Bhattacharjee
San Diego State University

Pearson Education, Inc.
Upper Saddle River, NJ 07458

Library of Congress Cataloging-in-Publication Data on file

Vice President and Editorial Director, ECS: *Marcia Horton*
Acquisitions Editor: *Laura Fischer*
Editorial Assistant: *Erin Katchmar*
Vice President and Director of Production and Manufacturing, ESM: *David W. Riccardi*
Executive Managing Editor: *Vince O'Brien*
Managing Editor: *David A. George*
Production Editor: *Scott Disanno*
Electronic Composition: *Clara Bartunek*
Cover Designer: *Bruce Kenselaar*
Art Editor: *Greg Dulles*
Manufacturin Manager: *Trudy Pisciotti*
Manufacturing Buyer: *Lisa McDowell*
Marketing Manager: *Holly Stark*

© 2003 Pearson Education, Inc.
Pearson Education, Inc.
Upper Saddle River, N.J. 07458

PRINTED IN THE UNITED STATES OF AMERICA

10 9 8 7 6 5 4 3 2

ISBN 0-13-009235-5

Pearson Education Ltd., *London*
Pearson Education Australia Pty. Ltd., *Sydney*
Pearson Education Singapore, Pte. Ltd.
Pearson Education North Asia Ltd., *Hong Kong*
Pearson Education Canada, Inc., *Toronto*
Pearson Educacíon de Mexico, S.A. de C.V.
Pearson Education—Japan, *Tokyo*
Pearson Education Malaysia, Pte. Ltd.
Pearson Education, Inc., *Upper Saddle River, New Jersey*

Contents

Preface

■ The History of TEST

I fell in love with Thermodynamics the day I was introduced to it during my undergraduate days. After starting to teach this subject, first as a graduate student at Washington State and then for several years as a professor at San Diego State University, I was struck by the fact that the majority of my students, including the ones who did well in the tests, were not grasping the concepts I was trying so hard to get across—the subject I loved was feared by most. I kept on changing my teaching style until I realized that it was not me, but the multitude of forms the governing equations assume, the huge number of apparently unrelated tables and charts, rapidly changing topics from lecture to lecture—some of the very things that make thermodynamics so general, so beautiful—are at the core of the problem. Despite my sustained effort to emphasize the central role the governing balance equations play to unite all the diverse topics, the favorite question from the students remained, "*Are we supposed to interpolate the tables during exams?*"

Soon after the advent of the web, I designed a web page for simplifying a thermodynamic system in a hierarchical tree structure. Starting from a general system, the students could click on a particular branch, say, open versus closed system, and the modified set of balance equations would be displayed. It was a huge hit with my students: they could click their way to arrive at the right set of governing equations for any given problem provided they learned how to ask the right question to simplify a system. However, there was no way of getting around interpolation of tabular data unless one was willing to write a program, something even the most dedicated students did not enjoy doing for the sake of learning thermodynamics. If only the property tables could be manipulated within a web page, the combination of the governing equations and a smart state finder could take the misery out of any solution. *Problem solving could be fun!*

Almost as an answer to my quest for a web-based tool, Sun Microsystems released Java. I quickly set out to learn object-orient programming in Java. What resulted after four years of work is TEST, The Expert System For Thermodynamics.

■ What is TEST?

TEST is a visual environment for solving problems, pursuing what-if scenarios, conducting numerical experiments and learning thermodynamics in a painless manner. The smart environment is created by a web of carefully designed HTML pages containing schematics, equations, pedagogic discussions, visual examples, and embedded Java applets, called the **daemons**. The daemons are smart thermodynamic calculators. The word daemon is not meant to scare away students from thermodynamics; a subject already suffering from more than its share of misunderstanding. It is more like a genie, a cross between Maxwell's thermodynamic demon and the silent background applications found in the Unix operating system.

Briefly, here is how TEST works. A problem-solving session begins by treating the system under consideration as the most general system with all possible interactions with the surroundings, a super system from which all other systems can be derived through simplification. Using a sequence of simplification tables, a user charts a course by selecting an appropriate branch (open versus closed, for instance) that best describes the problem. As more specificity is added to the system (steady versus unsteady, for instance), the changes in the system schematic and governing equations (mass, energy, and entropy balance equations) are displayed. The final table in this sequence offers a list of models for the working substance. As the user selects a particular idealization (say, the ideal gas model), the daemon, aware of the material properties and simplified governing equations customized for this particular problem, is launched. (This takes from 10 to 60 seconds over the Internet and less than two seconds when locally installed.)

The fun begins after a successful solution is obtained (yes, TEST will check to see if all the balance equations are satisfied by the answers generated). Because all the variables are visually exposed, one can study any conceivable what-if scenario with ease, by simply changing one or more variables and updating all calculations by a single click of a button (called Super-Calculate). A simple problem of finding the mass of water vapor in a room can be converted to a humidifier design exercise by studying the water requirement to achieve a desired relative humidity. A study of how condenser pressure affects the thermal efficiency of a Rankine cycle can be instantly turned around to see the effect of boiler pressure, or the turbine efficiency, or the maximum temperature—all without a single line of programming. It is not only the variables but the working fluid (what if R-134a is used instead of R-12?), the unit system (SI or English) or even the material model (what if cp is assumed constant?), that can be treated as parameters. I find this feature especially useful to back up theoretical conclusions with a quick in-your-face confirmation right in the classroom.

TEST does not require any installation; it is simply copied to your hard disk or a web server. The daemons, written 100% in Java, are browsed using a Java-enabled browser such as IE or Netscape Navigator. As a result, TEST can run on any platform (Windows, Unix, Linux, or MacOS) capable of running a Java-enabled browser. The inherent protections used by modern browsers shield TEST from crashing your system or passing on a virus. The other side of this security feature is that TEST cannot write to your disk directly. To save the TEST-Codes, you have to copy the code to a word processor (WordPad, for instance) first.

The Educational version of TEST is freely distributed to educators. Mirroring is encouraged so that everyone accesses the same version. In classrooms fitted with digital projectors, TEST can be used off the net without having to install it. A large number of universities around the world have installed and mirrored TEST and the installation base continues to grow rapidly. It is very possible that no matter where you are in the world, you will find TEST at your fingertips as long as you have an access to a web browser.

■ Integrating TEST with Textbooks

The introduction of electronic calculators almost instantly changed the paradigm of engineering education. The effect of the information revolution, however, is being selectively felt in different disciplines with thermodynamics education barely different today than it was 20 years ago. One only has to compare a modern thermodynamic textbook with an earlier edition to realize that the only thing that has changed is the addition of "computer problems" as an afterthought that are all but neglected by many educators due to lack of time. The installation of software, its maintenance, its dependence on a particular operating system—all these factors contribute into poor accessibility on the part of the students.

Being a platform-independent web-based software with a calculator-like graphical user interface, TEST can be relied upon like a calculator. Instead of adding computer problems at the end, it is now possible to use a computer to simplify a calculation just like calculators once did. A TEST solution can be used after every manual solution to double-check an answer (which, by itself, is a sound engineering practice) and then parametrically study the problem for gaining insight. It can be used as a numerical laboratory to explore the behavior of thermodynamic properties. A computer solution should aid, simplify and enhance a manual solution—not put additional burden on students. Designed to emulate the manual solution, TEST reinforces the classical approach to thermodynamic problem solving by engaging its user visually through the fundamental steps—simplification of the system, idealization of the working fluid, approximations, use of thermodynamic plots, and interpreting the answers—to obtain a solution without a single line of programming.

Integrating TEST to an existing textbook, however, is not a trivial task. After several semesters of trial and error, I have finally succeeded in charting a path through the existing textbooks that exploits the power of TEST and truly integrates computerized and manual solutions. I have begun working on a textbook based on this hybrid approach, which, when completed, will help educators in their effort to simplify and enhance thermodynamic education through the effective use of software. The progress on this project and other relevant discussions will be regularly posted on the Prentice Hall web site www.prenhall.com/thermo.

■ Using TEST in THE CLASSROOM

At SDSU, several classrooms are now equipped with digital projection systems connected to the Internet. Projecting TEST in the classroom as a teaching aid has never been easier. Students comments indicate that they cannot get enough of it in the classroom. Their only regret is they are not allowed to access TEST during a test. That day is not far off, when every student desk will be equipped with a networked device allowing access to TEST just like a calculator. Until that day arrives, a hybrid approach where TEST complements the traditional style seems to be the best answer.

I introduce TEST to each class during the very first week as a numerical laboratory; students start exploring the behavior of different properties as they are introduced in the lectures—how, for instance, the internal energy or specific volume of a substance changes with temperature and pressure. Numerical evaluation of entropy is no more difficult than the evaluation of specific volume, so why not use it to grow an appreciation for entropy as a measure of disorder from the very beginning?

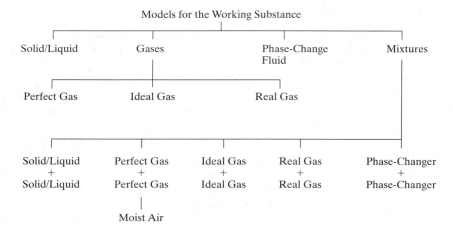

■ **Figure 1** Classification of models used to idealize a working substance.

Instead of the traditional approach of starting with ideal gas and then gradually introducing other models, I start with a complete map of the material models, as shown in Figure 1. Through numerical experiments, the students compare different models and learn when to use which. A comparison among the real gas, ideal gas and perfect gas model, for instance, underscores the trade-off between accuracy and simplicity.

While the traditional manual evaluation of state is thoroughly covered, life is made easier by limiting complete manual solutions to the following materials: (a) solid and liquid modeled by the solid–liquid model, (b) gases modeled by the perfect gas model, and (c) steam and R-134a as representative phase-change fluids. The time saved by delegating problems involving complex models to TEST is utilized in learning when to use such models. The variable specific heat assumption that distinguishes the ideal gas model from the perfect gas model can be explored without the pain of interpolating artifical variables such as the relative pressure or relative volume. Combination of a manual solution with the perfect gas assumption and the TEST solution with different models and different fluids provide a much more interesting alternative with a deeper insight. While a manual solution mostly focuses on how to obtain an answer, a TEST solution develops the skill to judge the goodness of a solution. Every state evaluation is automatically accompanied by a thermodynamic plot (unless that feature is turned off). Thermodynamic relations such as the T-ds equation take on new meaning as the students apply them to test a database, not just to derive a formula.

There are two types of state daemons. Besides calculating the equilibrium state, the *Volume State* daemons are used to evaluate the total amount of mass, energy, or entropy of a control mass. The *Surface State* daemons, on the other hand, are used to evaluate how a flow carries properties such as energy or entropy. Numerical evaluation of the total amount of property in a control volume and the flow rate of the property across the control surface leads to the development of the balance equations for mass, energy, entropy, and exergy.

The generic systems, typically encountered in the first half of most textbooks (Chapters 3–7 of Archive), are classified according to the degree of simplification, and the corresponding generic system daemons are arranged in a tree structure as shown in Figure 2. In general, most system analysis involves identification of system boundary, simplification of the system, customizing the balance equa-

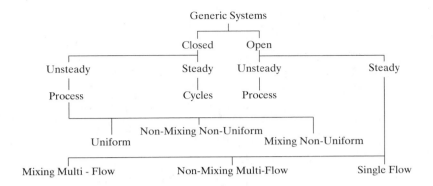

Figure 2 Classification of a generic system. Any thermodynamic problem from the first half of most thermodynamic texts can be simplified into one of these categories.

tions, idealization of the working substance, analysis of state and governing balance equations, verification of results, and parametric study.

Instead of spending an entire class on a single comprehensive manual solution of a system problem, I use batch processing; that is, solving a number of problems of similar kinds one step at a time. Just the simple exercise of locating the right system daemon for the problem forces a user to use step-by-step simplification, verify the balance equation, and idealize the working substance through the choice of a material model. A problem involving energy balance on a piston-cylinder device is treated as a "closed-system" problem rather than an "energy balance" problem. Once one learns to locate the appropriate daemon for this problem, all balance equations, including exergy, are displayed on their customized format.

The system daemons build upon the state daemons by adding an analysis panel that solves the customized balance equations. The solution of an energy problem also produces solution for the entropy and, with a little additional effort, exergy equations. A TEST solution for a problem involving boundary work evaluation also produces the entropy generation, the heat transfer, and all other relevant variables. Such a solution cannot only be used to verify a manual solution, but can be used to explore how the temperature increases during adiabatic compression is affected by the compression ratio or any other parameter. Simply change a variable and click the Super-Calculate button to update all calculations. It is that simple. Further, by copying the TEST-Codes among different models, the difference among perfect gas, ideal gas, real gas, or a phase-change model can be quickly explored.

The frequently used daemon for closed systems is definitely the *Systems.Closed.Process.Generic.Uniform* daemon. The open-system problems are similarly covered in a comprehensive manner with the *Systems.Open.SteadyState.Generic.Singleflow* daemon being the most utilized open-system daemon.

One of the significant aspects of the TEST solution is that it aids in explaining the fundamental concepts through numerical experiments. The generation of entropy during mixing, for instance, can be explored through the *Open.SteadyState.Generic.MultiflowMixing* daemon. All that is required is the evaluation of three surface states. By changing the species and the inlet properties a lot can be learned regarding entropy generation than by manipulating equations alone. Students today are much more numerically inclined and readily absorb the concepts.

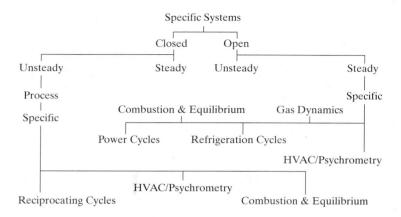

■ **Figure 3** Classification of application-specific systems. Although a specific system (HVAC, for instance) can fall into two logical groups, open-steady or closed-process, it is treated in a single chapter in textbooks.

The second half of most texts (Chapters 8–15 in the Archive) deals with application-specific systems and TEST organizes them in a structure (Figure 3) similar to the generic systems. The resulting daemons, called the *specific daemons*, build upon the generic daemons by adding domain-specific panels—reaction panel for the combustion daemon, cycle panel for the cycle daemons, table panel for the gas dynamics daemon—to the familiar states and analysis panels. Merging Figure 2 with Figure 3, the organizational structure of all the TEST daemons is created in what is called the TEST-Map. An advanced user can avoid the step-by-step simplification procedure and directly launch a desired daemon through the use of this map.

Some of the specific daemons go beyond what is typically covered in an undergraduate course. The HVAC daemon does not have the same limitations of the standard psychrometric chart and can handle dew point temperatures below freezing or a total pressure different from one atmosphere. The refrigeration daemons contain refrigerant mixtures identified with the percent suffix (the vapor pressure of such mixtures varies with quality), which are finding increasing use in professional applications. The gas dynamics daemon offers smart tables for oblique shocks and Prandtl–Meyer expansion wave in addition to the standard isentropic and normal shock tables.

I use the specific daemons in the second course of thermodynamics, a course with significant design content. Almost every problem we solve in the class is turned around into a "*what-if ?*" study, the hallmark of good design analysis. As with the generic systems, basic problems are first covered using the side-by-side manual and TEST analysis. Once the students are confident regarding obtaining accurate solutions, the parametric studies begin. Almost every homework problem involves such studies, with the students picking their own parameter of choice in a given problem. The TEST-Codes found in the Archive come quite handy in this regard. Because most of the basic cycles or fundamental processes in HVAC can be jump-started using the TEST-Code (avoiding what can be a lengthy session of entering a number of states, analyzing a number of devices or process and then combining them into a cycle for instance), parametric studies in the classroom becomes quite convenient. The qualitative conclusions such as how reheating can affect the thermal efficiency, can be quickly backed up empirically.

To what extent TEST should be integrated with teaching? It all depends on the personal style of an educator and the state of the art of the web access. Perhaps the following comment from one of my students best describes the infancy of software use in thermodynamic education. He asked, and I paraphrase, "I am quite confident about solving almost any thermodynamic problem with TEST on my side. Without TEST, however, I am much slower and am never sure if I got the right answer. Does this mean that I have not learned thermodynamics well?" Just like a calculator, TEST can be relied upon to be accessible anywhere in the world—embracing a tool such as TEST is no different than evaluating a trigonometric function using a calculator.

ACKNOWLEDGMENTS

TEST would not be possible without the feedback I have received from thousands of users reporting bugs, suggesting improvements, and expressing support and encouragement, and, many times, engaging me in exciting pedagogical exchanges. Several students and colleagues have helped me throughout the development of TEST. Wonchul Jung, Chuck Parme, Shan Liang, Matt King, Jeff West, Guy Fujiwara, and Albert Nguyen are some of the students I must single out. I am indebted to my colleagues Professor Roger Whitney, Bob Cademy, Mark Boyns, and Andrew Scherpbier, who helped me learn object-oriented programming, the late Preston Lowrey and George Craig for countless thermodynamic discussions over coffee, and to Chris Paolini for managing the web servers at SDSU. Special credit is due to Tanuka Bhattacharjee, who designed the colors used for highlighting the daemons. The profuse use of the colors blue, green, and red everywhere in TEST, however, was dictated by my children, Robi, Sarah, and Neil, who are a constant source of joy in my life. I am enormously grateful to Kyoung (Keya) Bhattacharjee for helping me with the data structure, program design, and for being my wife.

Finally, I would like to express my sincere gratitude to Tim Bozik at Prentice Hall, who instantly recognized TEST's potential as a learning tool, to Laura Fischer, the acquisitions editor, who worked with me from the beginning to end of the manuscript preparation, and Holly Stark, Scott Disanno, and David George, who worked hard to make this project a reality. And last but not the least, I thank Sun Microsystems for supporting part of this project through an Academic Grant and, more importantly, for inventing the Java programming language.

SUBRATA (SOOBY) BHATTACHARJEE
San Diego State University

Visual Tour

TEST uses an intuitive graphical user interface in creating a visual environment that requires no learning curve to master. However, there are some features, which, if gone over in a systematic manner, can make problem solving quite rewarding. For instance, moving the pointer over the selected substance brings up additional information on the Message Panel, or using the Input/Output panel as a full-blown scientific calculator that recognizes property symbols and adheres to familiar Microsoft Excel syntax are some of the features that are not immediately apparent. And then there are those subtle features that should be best left for the users to unearth. Like this student posted on the TEST Comments box, "Everyday, I discover new secrets in TEST."

This chapter presents a tour of the entire site with emphasis on the state daemons. It is recommended that you read this chapter with TEST open on your computer screen, where you can experiment with some of the daemons highlighted in this chapter. The complete distribution of the academic edition of TEST is included in the attached CD. For the best performance, install TEST in your computer following the simple instructions on the README file in the CD. To run TEST from the CD directly, click on the index.html file, provided you have a relatively recent version of Netscape or Internet Explorer. If you have good Internet connection, you can also launch the latest edition of TEST over the web from www.prenhall.com/thermo.

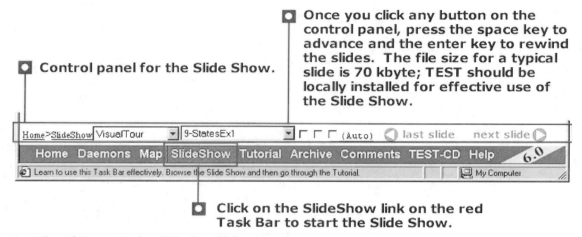

Control panel for the Slide Show.

Once you click any button on the control panel, press the space key to advance and the enter key to rewind the slides. The file size for a typical slide is 70 kbyte; TEST should be locally installed for effective use of the Slide Show.

Click on the SlideShow link on the red Task Bar to start the Slide Show.

Visual Tour: Navigating the Slide Show

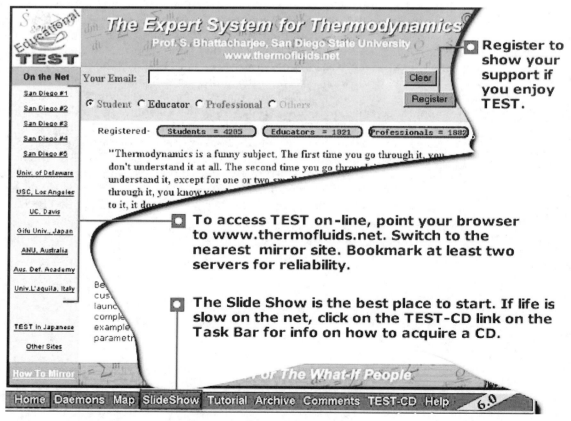

Register to show your support if you enjoy TEST.

To access TEST on-line, point your browser to www.thermofluids.net. Switch to the nearest mirror site. Bookmark at least two servers for reliability.

The Slide Show is the best place to start. If life is slow on the net, click on the TEST-CD link on the Task Bar for info on how to acquire a CD.

Visual Tour: Start at the TEST Home Page

- To install **TEST** in your PC (running Windows, Unix or MacOS):

  a) Acquire a **TEST**-CD (free for educators).

  b) Copy **TEST** into your hard drive.

  c) Install Navigator (4+) or IE (5+) if your browser is too old.

■ *Visual Tour*: **Installing TEST from a CD**

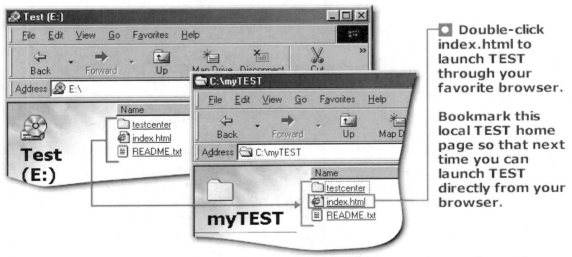

- Double-click index.html to launch **TEST** through your favorite browser.

  Bookmark this local **TEST** home page so that next time you can launch **TEST** directly from your browser.

- Create a folder — say, myTEST — on the hard drive, and copy the entire content of the CD into that folder. That is all that is necessary to install TEST — no DLL files, no exe files, no restart, and no blue screens.

■ *Visual Tour*: **Copying TEST into Hard Drive**

TEST is a general-purpose, visual, thermodynamic problem-solving environment for engineers and scientists, covering a wide range of topics — from thermodynamic state evaluation through gas dynamics. Started as Webware for undergraduate thermo classes, TEST has rapidly evolved into a comprehensive software package used by professionals and students alike.

Briefly, this is how TEST works. The problem-solving session begins with the assumption of the most general thermodynamic system. As the user simplifies the problem by charting a suitable path (open vs. closed system, for example) through a series of truth tables, the system schematic, and the balance equations adjust to each decision. Finally, an idealized model for the working fluid is chosen and a Java application, called a daemon, knowledgeable about material properties and the customized set of balance equations, is launched. An intuitive graphical user interface makes it possible to solve the problem on-line without a single line of programming.

However, the fun begins after a problem is solved. Because all the variables of the problem are visually exposed, any input value can be changed and its effect on the solution studied by a single click of the Super-Calculate button. This feature makes it extremely easy to pursue any conceivable what-if scenario.

The visual solution is accompanied by a spreadsheet-friendly property table, a printer-friendly detailed report and a few lines of solution macro called the TEST-Code. The original solution can be regenerated instantly by loading a previously save TEST-Code. This makes it easy to archive and share TEST solutions.

■ *Visual Tour*: **What is TEST?**

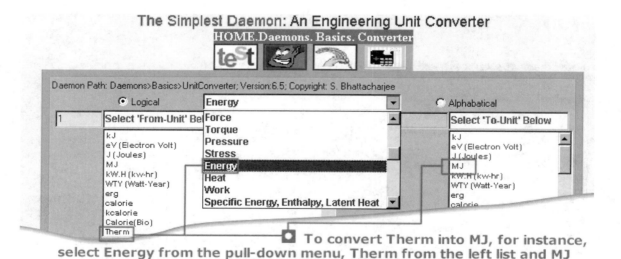

The Simplest Daemon: An Engineering Unit Converter

To convert Therm into MJ, for instance, select Energy from the pull-down menu, Therm from the left list and MJ from the right. The daemon displays 1 Therm = 105.5 MJ.

■ *Visual Tour*: **The Simplist Daemon**

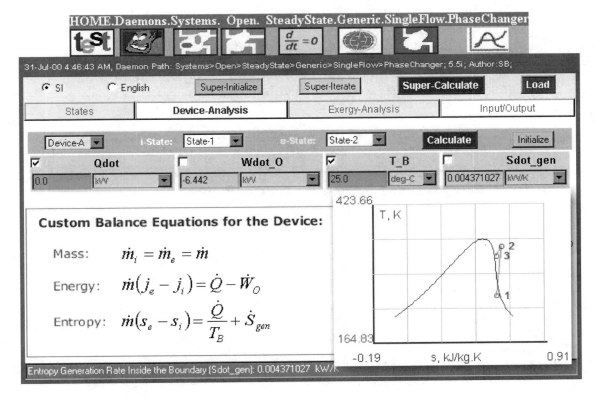

Visual Tour: Image of a Complex Daemon

The Message Panel is a constant feature in all daemons. To use it as a scientific calculator, move the pointer into the panel, type in an expression using Microsoft Excel syntax and press the Enter key.

Visual Tour: The Task Bar

● State:

In **TEST** a State is defined as a snapshot of variables (thermodynamic, extrinsic, system, and material properties) that describe the local equilibrium of a system at a given instant at a given location within the control volume.

The State Daemon is the basic building block of all other daemons. The next few slides highlight some of its features that are shared by all other daemons.

■   *Visual Tour*: **The State Daemons, The Building Block**

Example: Determine the mass of steam in a tank of volume 500 gallons for the following conditions: p=100 kPa; Quality=50%. What fraction of the volume does the saturated vapor occupy?

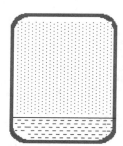

▣ Solution Procedure: Simplify the problem as a volume state problem. Model the working fluid as a phase-change fluid and select H2O. Enter the known variables and Calculate the rest.

■   *Visual Tour*: State Evaluation Example #1

The TEST Algorithm:

- Simplify the problem using the TEST-Map or the systematic step-by-step procedure starting with the Home.Daemons page.

- Pick an appropriate material model (Idealize).

- For each combination, a unique daemon, a customized tool for the problem, appears on the main window.

■ *Visual Tour*: Daemon – The Customized Thermodynamic Calculator

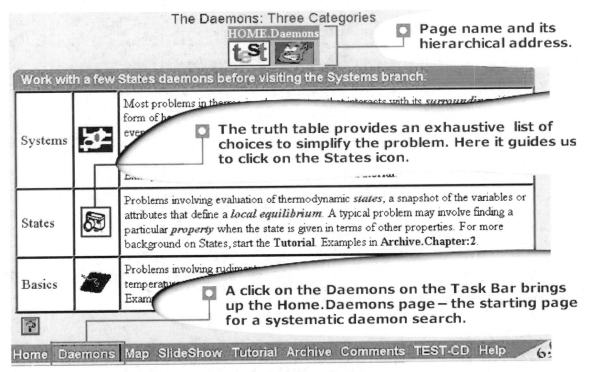

The Daemons: Three Categories

HOME.Daemons

Page name and its hierarchical address.

Work with a few States daemons before visiting the Systems branch.

Systems		Most problems in the... form of h... ever...
States		Problems involving evaluation of thermodynamic *states*, a snapshot of the variables or attributes that define a *local equilibrium*. A typical problem may involve finding a particular *property* when the state is given in terms of other properties. For more background on States, start the **Tutorial**. Examples in **Archive.Chapter:2**.
Basics		Problems involving rudiment... temperatur... Exam...

The truth table provides an exhaustive list of choices to simplify the problem. Here it guides us to click on the States icon.

A click on the Daemons on the Task Bar brings up the Home.Daemons page – the starting page for a systematic daemon search.

Home   Daemons   Map   SlideShow   Tutorial   Archive   Comments   TEST-CD   Help   6:

■ *Visual Tour*: Step-by-Step Simplification Starts Here

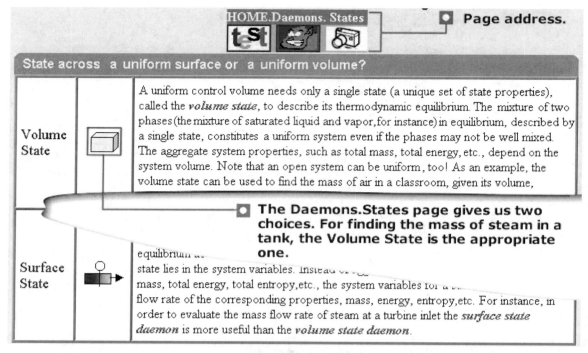

Page address.

State across a uniform surface or a uniform volume?

Volume State

A uniform control volume needs only a single state (a unique set of state properties), called the *volume state*, to describe its thermodynamic equilibrium. The mixture of two phases (the mixture of saturated liquid and vapor, for instance) in equilibrium, described by a single state, constitutes a uniform system even if the phases may not be well mixed. The aggregate system properties, such as total mass, total energy, etc., depend on the system volume. Note that an open system can be uniform, too! As an example, the volume state can be used to find the mass of air in a classroom, given its volume,

The Daemons.States page gives us two choices. For finding the mass of steam in a tank, the Volume State is the appropriate one.

Surface State

equilibrium ...
state lies in the system variables. Instead of ...
mass, total energy, total entropy, etc., the system variables for a ...
flow rate of the corresponding properties, mass, energy, entropy, etc. For instance, in order to evaluate the mass flow rate of steam at a turbine inlet the *surface state daemon* is more useful than the *volume state daemon*.

■ *Visual Tour*: **Step-by-Step Simplification – The Daemons. States Page**

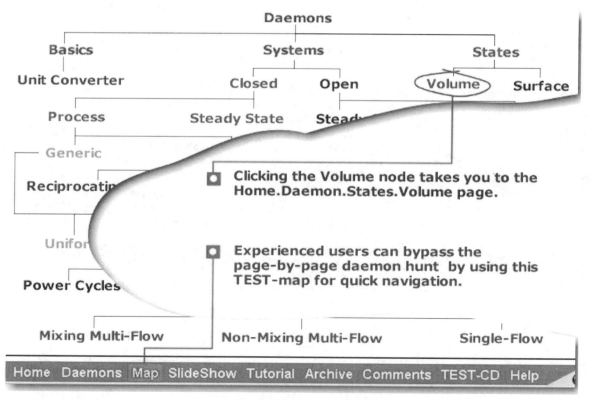

Daemons

Basics   Systems   States

Unit Converter   Closed   Open   Volume   Surface

Process   Steady State   Stead

Generic

Reciprocatin

Clicking the Volume node takes you to the Home.Daemon.States.Volume page.

Unifor

Experienced users can bypass the page-by-page daemon hunt by using this TEST-map for quick navigation.

Power Cycles

Mixing Multi-Flow   Non-Mixing Multi-Flow   Single-Flow

Home Daemons Map SlideShow Tutorial Archive Comments TEST-CD Help

■ *Visual Tour*: **Direct Navigation to the Daemon Page, Using the TEST-Map**

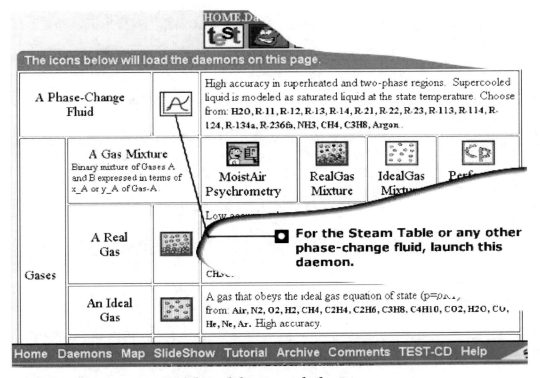

■ **Visual Tour**: Select a Material Model to Launch the Daemon

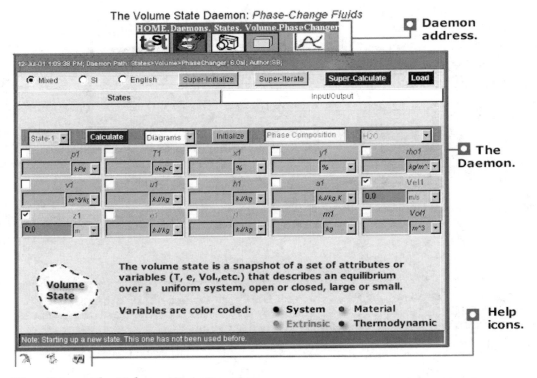

■ **Visual Tour**: The Volume State Daemon

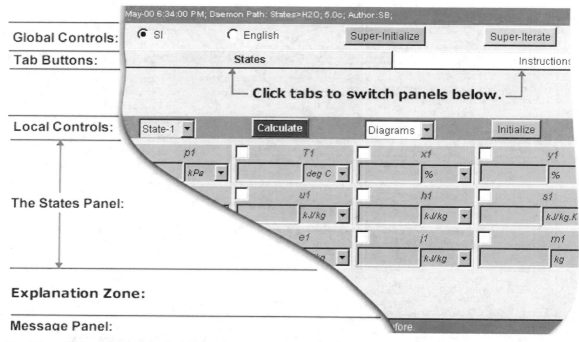

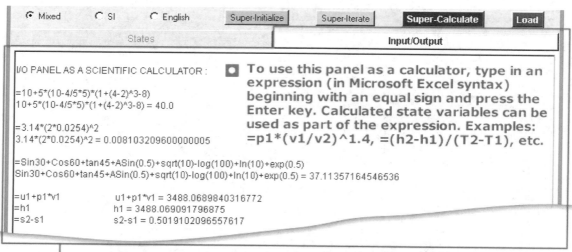

■ **Visual Tour: Layout of a State Daemon**

■ **Visual Tour: States Panel**

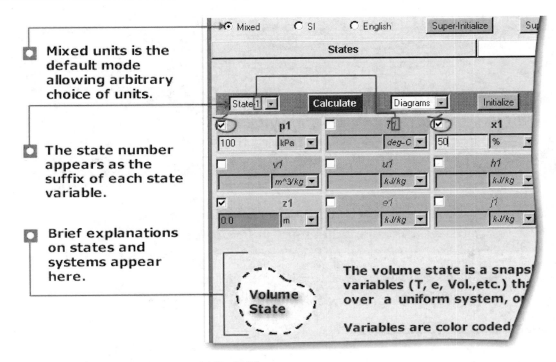

Mixed units is the default mode allowing arbitrary choice of units.

The state number appears as the suffix of each state variable.

Brief explanations on states and systems appear here.

■ *Visual Tour*: **Enter Known Variables**

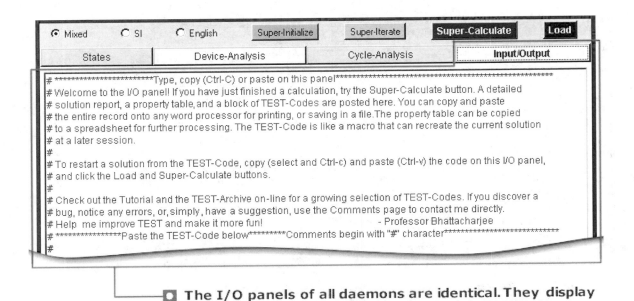

The I/O panels of all daemons are identical. They display the solution report, property table, and TEST-Code. They also accept TEST-Code to regenerate a solution.

■ *Visual Tour*: **I/O Panel**

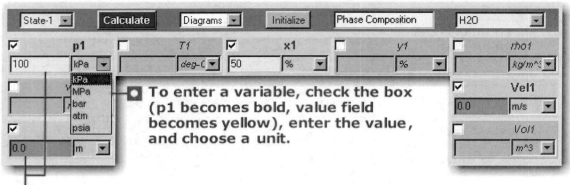

To enter a variable, check the box (p1 becomes bold, value field becomes yellow), enter the value, and choose a unit.

The value field assumes different background colors:

- Yellow: Editable values not yet read by the daemon.
- Green: Default or entered values that have been read.
- Cyan: Values calculated by the daemon.
- White: Failure to evaluate or read the value.

■ *Visual Tour*: **Background Colors**

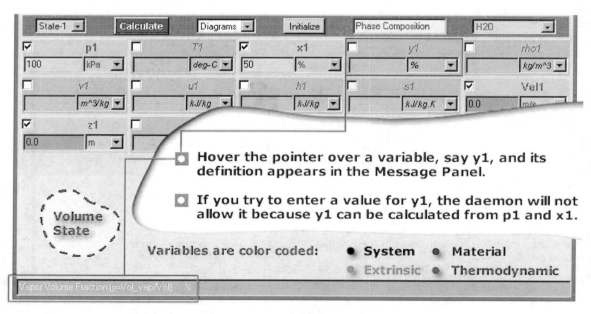

Hover the pointer over a variable, say y1, and its definition appears in the Message Panel.

If you try to enter a value for y1, the daemon will not allow it because y1 can be calculated from p1 and x1.

Variables are color coded: ● **System**   ● Material
● Extrinsic   ● Thermodynamic

■ *Visual Tour*: **The Daemon Guards against Redundant Input**

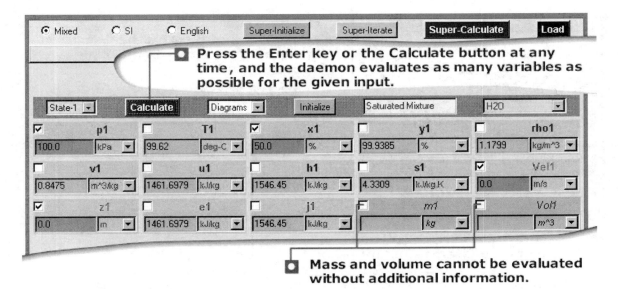

Press the Enter key or the Calculate button at any time, and the daemon evaluates as many variables as possible for the given input.

Mass and volume cannot be evaluated without additional information.

■ *Visual Tour*: **Calculate the State**

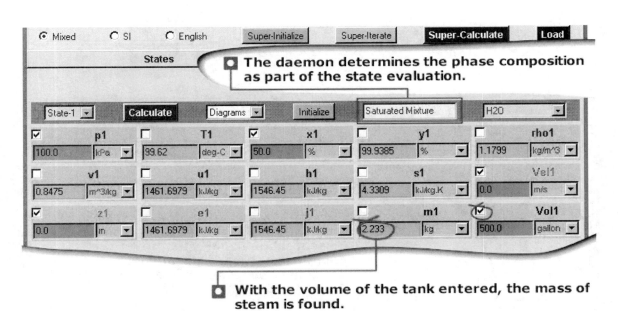

The daemon determines the phase composition as part of the state evaluation.

With the volume of the tank entered, the mass of steam is found.

■ *Visual Tour*: **Further Calculations—System Variables m and Vol**

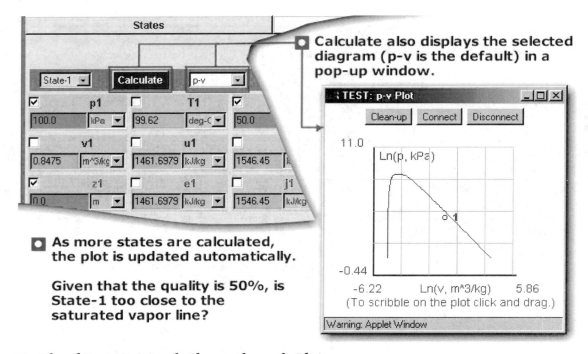

● Calculate also displays the selected diagram (p-v is the default) in a pop-up window.

● As more states are calculated, the plot is updated automatically.

Given that the quality is 50%, is State-1 too close to the saturated vapor line?

■ *Visual Tour*: **Automatic Thermodynamic Plot**

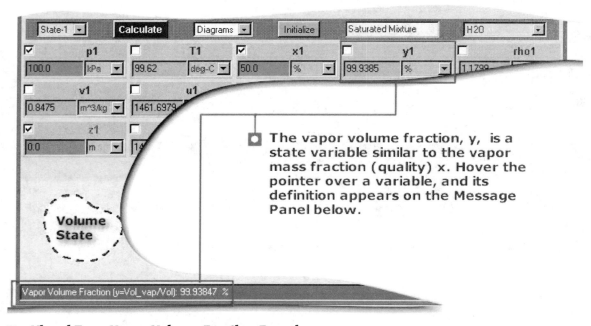

● The vapor volume fraction, y, is a state variable similar to the vapor mass fraction (quality) x. Hover the pointer over a variable, and its definition appears on the Message Panel below.

■ *Visual Tour*: **Vapor Volume Fraction Found**

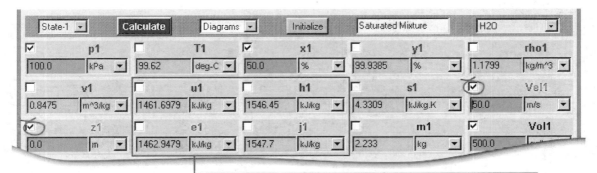

Variables Vel and z are initialized to zero because they have only slight effects on e and j, defined as

$$e \equiv u + \frac{Vel^2}{2} + gz; \quad j \equiv h + \frac{Vel^2}{2} + gz$$

While e, the specific total energy, is a standard symbol, we introduce the symbol j to represent the total specific enthalpy, also known as the specific flow energy. In open system analysis, j plays a crucial role in the energy balance equation.

■ *Visual Tour*: **Further Calculations—Extrinsic Variables e and j**

◼ The same state is found from another set of inputs: two thermo properties u and v, one system property m, and two extrinsic properties e and z.

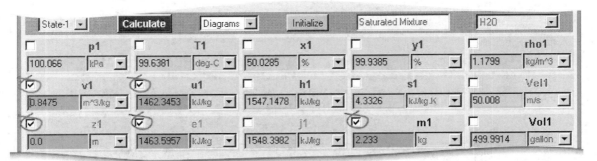

The state daemons accept any combination of input properties (after making sure that they are independent) and determine as many dependent properties as possible.

■ *Visual Tour*: **Different Input, Same State**

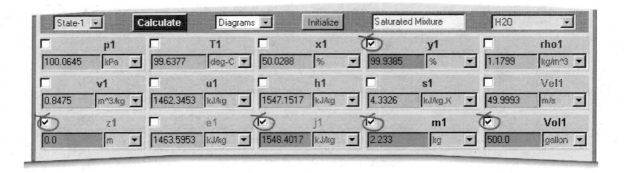

■ Here the same state is found from yet another set of inputs: two system properties, m and Vol, one thermo property, the vapor volume fraction y, and two extrinsic properties, j and z.

■ *Visual Tour*: **Different Input, Same State**

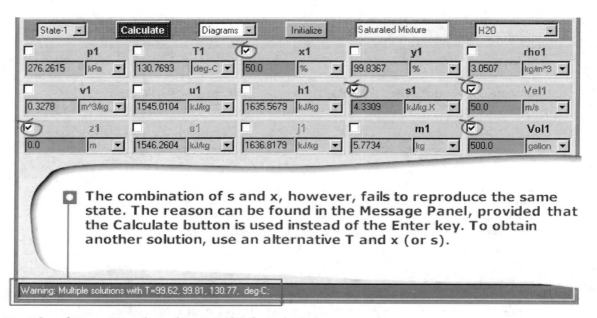

■ The combination of s and x, however, fails to reproduce the same state. The reason can be found in the Message Panel, provided that the Calculate button is used instead of the Enter key. To obtain another solution, use an alternative T and x (or s).

Warning: Multiple solutions with T=99.62, 99.81, 130.77, deg-C;

■ *Visual Tour*: **Warning about Multiple Solutions**

■ **Select State-2, enter p2=1 kPa and x2=x1. Calculate.**
　　■ **Note the use of Excel-type algebraic expression.**

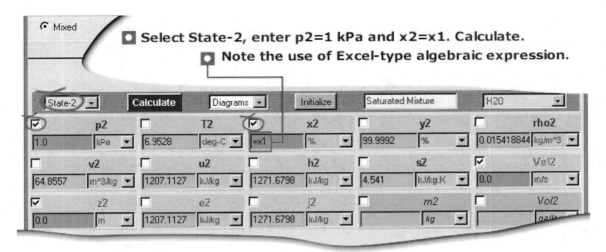

■ To explore the reason behind the multiple solutions seen in the
last slide, let us create a locus of x=50% on a T–s diagram.

■ **Visual Tour:** **x = 50% Line on the T–s Plot**

■ In evaluating an expression like p2+4, p2 is picked up
from State-2, converted to the local unit, and added
to 4. If p2 were 10 psia, then p3 would be
68.95+4=72.95 kPa. Here, of course, p3 is 100 kPa,
which can be displayed on the Message Panel by
hovering the pointer over the p3 widget.

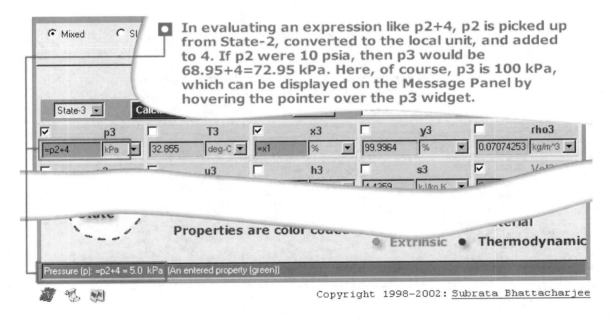

Copyright 1998-2002: <u>Subrata Bhattacharjee</u>

■ **Visual Tour:** **State-3, Evaluation of Expressions**

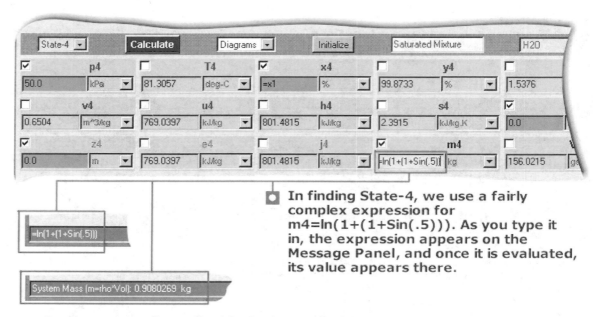

In finding State-4, we use a fairly complex expression for m4=ln(1+(1+Sin(.5))). As you type it in, the expression appears on the Message Panel, and once it is evaluated, its value appears there.

■ *Visual Tour*: **Use of Complex Algebraic Expressions (State-4)**

The S-shaped locus of x=50%, obviously, will yield multiple solutions for s=s1.

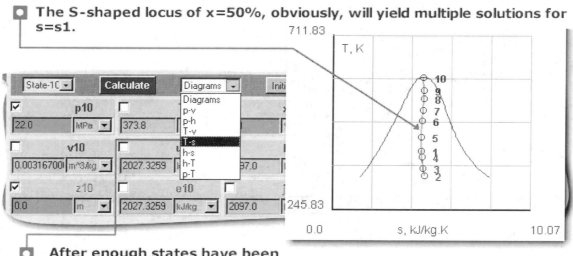

After enough states have been calculated, choose a suitable plot, T–s in this case, and the plot will appear on a floating window.

■ *Visual Tour*: **The Locus of x = 50%**

◘ The locus of x=50% in the previous slide begs a curious mind to explore whether such multiple solutions are possible at another quality, say, x=25%. The daemon offers a simple way to leverage a solution for what-if studies. Simply change x1 to 25% (press Enter) and click the Super-Calculate button.

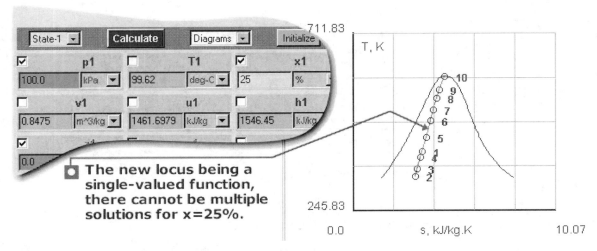

◘ The new locus being a single-valued function, there cannot be multiple solutions for x=25%.

■ *Visual Tour*: **What-If the Quality Is 25%?**

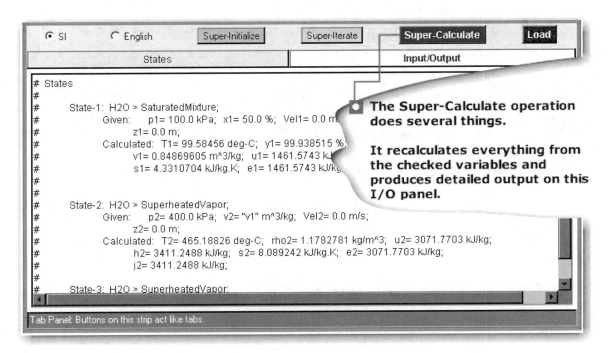

■ *Visual Tour*: **The Detailed Solution Report**

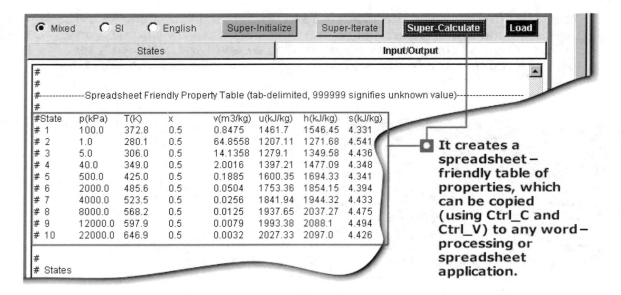

It creates a spreadsheet-friendly table of properties, which can be copied (using Ctrl_C and Ctrl_V) to any word-processing or spreadsheet application.

■ *Visual Tour*: **The Spreadsheet-Friendly Property Table**

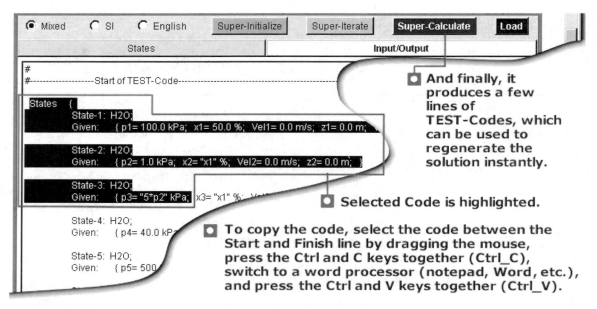

And finally, it produces a few lines of TEST-Codes, which can be used to regenerate the solution instantly.

Selected Code is highlighted.

To copy the code, select the code between the Start and Finish line by dragging the mouse, press the Ctrl and C keys together (Ctrl_C), switch to a word processor (notepad, Word, etc.), and press the Ctrl and V keys together (Ctrl_V).

■ *Visual Tour*: **The TEST-Code on the I/O Panel**

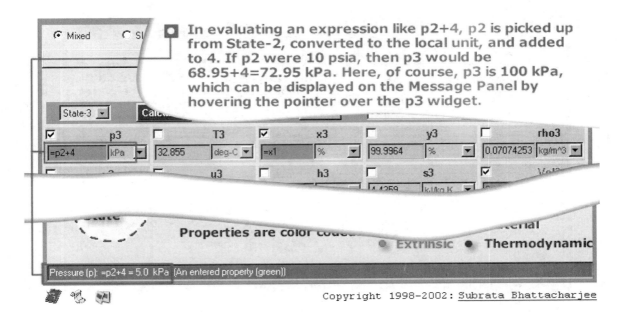

In evaluating an expression like p2+4, p2 is picked up from State-2, converted to the local unit, and added to 4. If p2 were 10 psia, then p3 would be 68.95+4=72.95 kPa. Here, of course, p3 is 100 kPa, which can be displayed on the Message Panel by hovering the pointer over the p3 widget.

Visual Tour: **Algebraic Expressions**

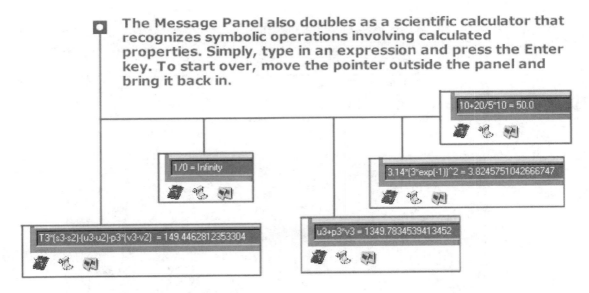

The Message Panel also doubles as a scientific calculator that recognizes symbolic operations involving calculated properties. Simply, type in an expression and press the Enter key. To start over, move the pointer outside the panel and bring it back in.

Visual Tour: **Message Panel**

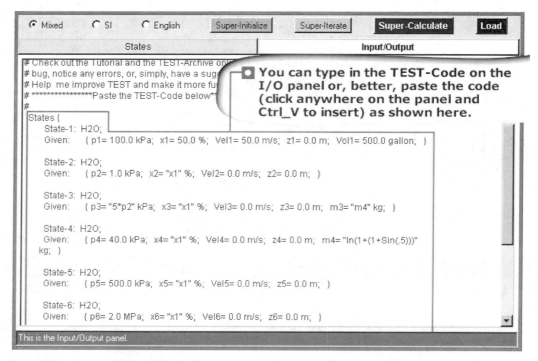

Visual Tour: Loading TEST-Code

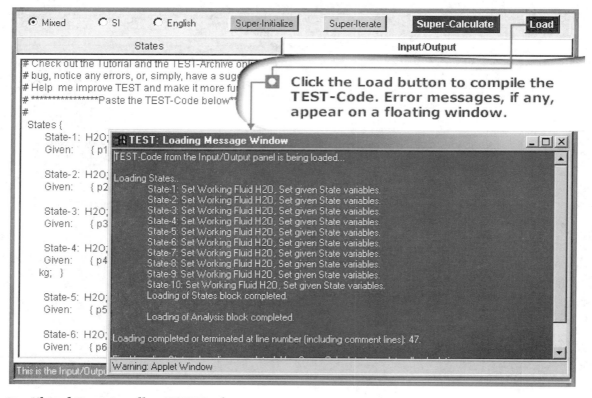

Visual Tour: Loading TEST-Code

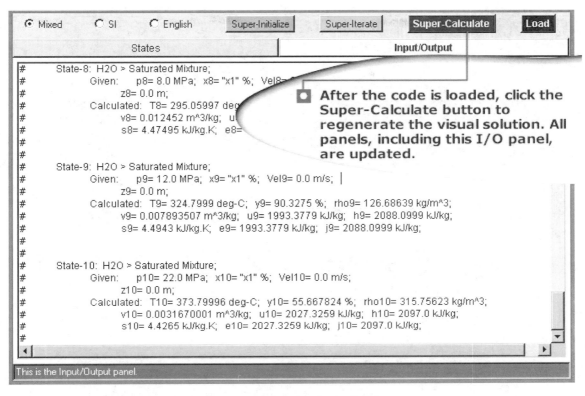

After the code is loaded, click the **Super-Calculate** button to regenerate the visual solution. All panels, including this I/O panel, are updated.

■ *Visual Tour*: **Super-Calculate to Regenerate the Solution**

What if we have R-134a in the tank instead of H2O? Simply pick the desired fluid from the pull-down menu and the new mass is calculated on the fly.

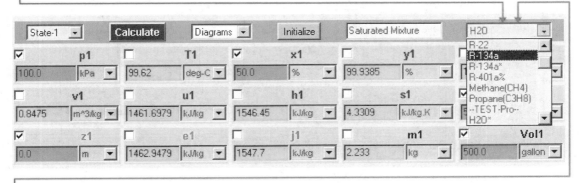

To obtain additional info on a material, hover the pointer over the species menu. Results are displayed on the Message Panel.

■ *Visual Tour*: **What-If Scenario—Change Working Fluid**

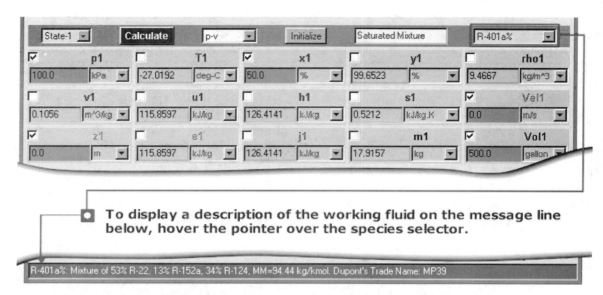

To display a description of the working fluid on the message line below, hover the pointer over the species selector.

R-401a%: Mixture of 53% R-22, 13% R-152a, 34% R-124, MM=94.44 kg/kmol. Dupont's Trade Name: MP39

■ *Visual Tour*: **Molar Mass and Other Information on the Working Fluid**

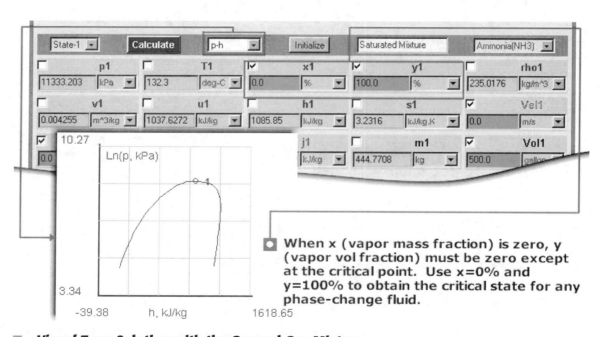

When x (vapor mass fraction) is zero, y (vapor vol fraction) must be zero except at the critical point. Use x=0% and y=100% to obtain the critical state for any phase-change fluid.

■ *Visual Tour*: **Solution with the General Gas Mixture**

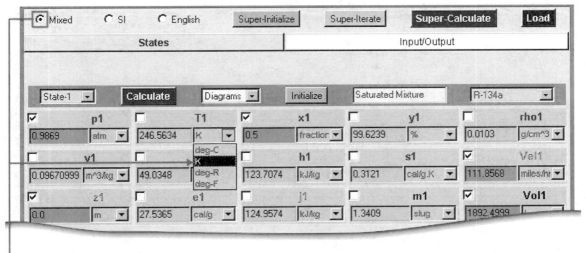

In the Mixed mode, any available unit can be chosen for individual variables. Choose a unit, and the value is adjusted on the fly.

■ *Visual Tour*: **Mixed Units**

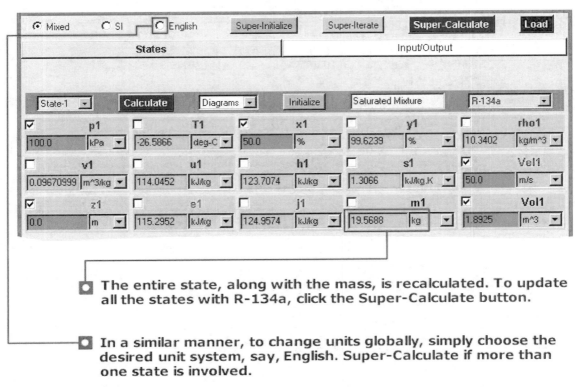

The entire state, along with the mass, is recalculated. To update all the states with R-134a, click the Super-Calculate button.

In a similar manner, to change units globally, simply choose the desired unit system, say, English. Super-Calculate if more than one state is involved.

■ *Visual Tour*: **Change Units**

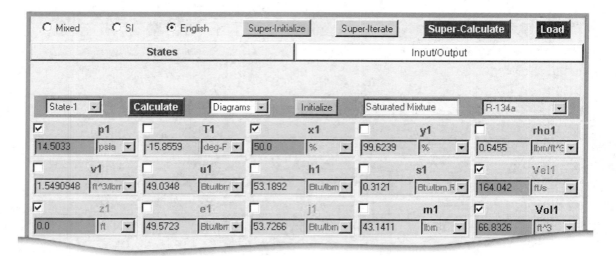

■ *Visual Tour*: English Units—The State in English System

Gas Models:

TEST treats all three gas models - perfect gas, ideal gas and real gas - in a uniform manner. Artifical variables such as relative volume, relative pressure, etc., are no longer necessary.

A binary gas mixture is created by mass-fraction weighted specific properties of any two gases. TEST also offers an ideal gas mixture model having up to 32 different constituents.

■ *Visual Tour*: Uniform Interface for All Gas Models

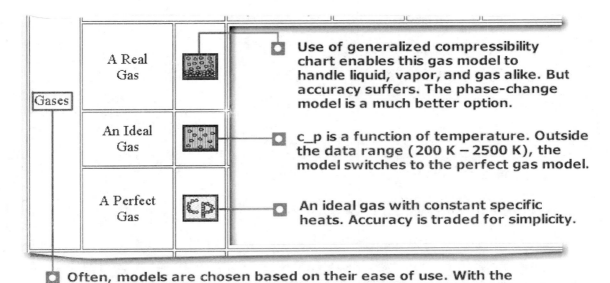

Gases

A Real Gas — Use of generalized compressibility chart enables this gas model to handle liquid, vapor, and gas alike. But accuracy suffers. The phase-change model is a much better option.

An Ideal Gas — c_p is a function of temperature. Outside the data range (200 K – 2500 K), the model switches to the perfect gas model.

A Perfect Gas — An ideal gas with constant specific heats. Accuracy is traded for simplicity.

Often, models are chosen based on their ease of use. With the standardized GUI of a state daemon, model selection can be based entirely on physics rather than convenience.

■ *Visual Tour*: **The Three Gas Models in TEST**

Example: Air at 40 psia and 50 deg-F flows through a duct of area 72 sq. in. with a velocity of 30 ft/s. Determine the mass flow rate in kg/min.

Treat air as (a) a perfect gas; (b) an ideal gas; (c) a real gas; (d) an ideal gas mixture of oxygen (23% by mass) and nitrogen.

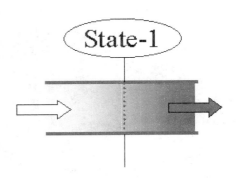

State-1

Solution Procedure: Simplify the problem as a surface state problem. Select the appropriate gas model to launch the daemon. Enter known variables. Calculate to find the complete state, including the mass flow rate.

■ *Visual Tour*: **A Simple Problem Involving Surface State**

The Surface State Daemon: *Perfect Gas*

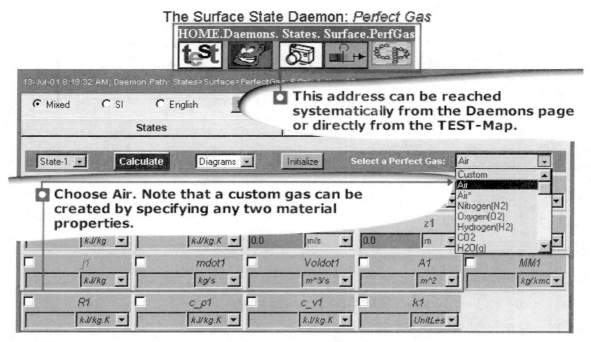

This address can be reached systematically from the Daemons page or directly from the TEST-Map.

Choose Air. Note that a custom gas can be created by specifying any two material properties.

■ *Visual Tour*: Launch the Perfect Gas Surface Daemona

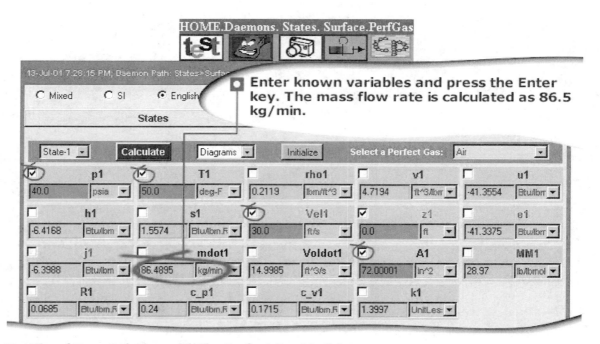

Enter known variables and press the Enter key. The mass flow rate is calculated as 86.5 kg/min.

■ *Visual Tour*: Solution with the Perfect Gas Model

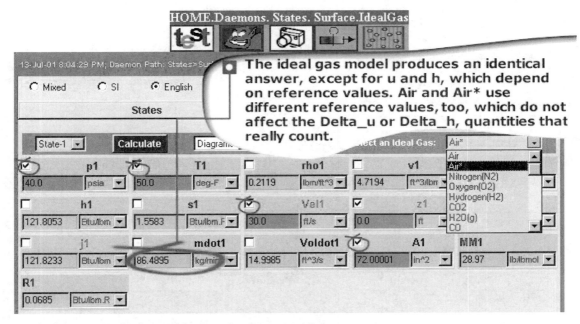

Visual Tour: **Solution with the Ideal Gas Model**

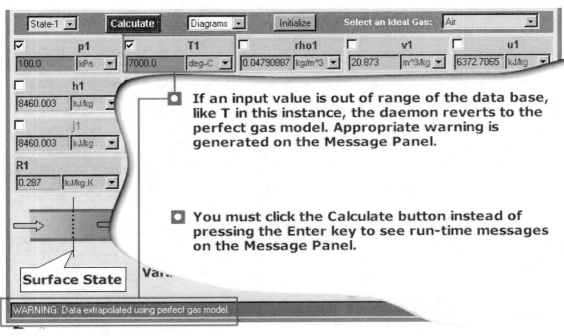

Visual Tour: **Data Out of Range**

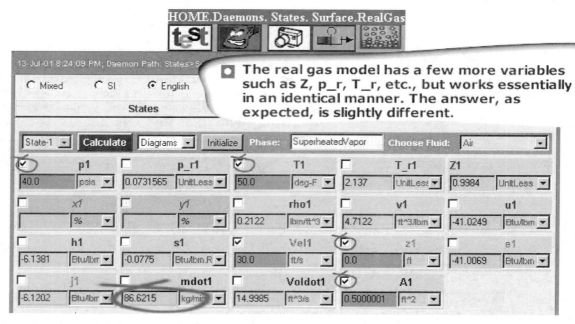

Visual Tour: Solution with the Real Gas Model

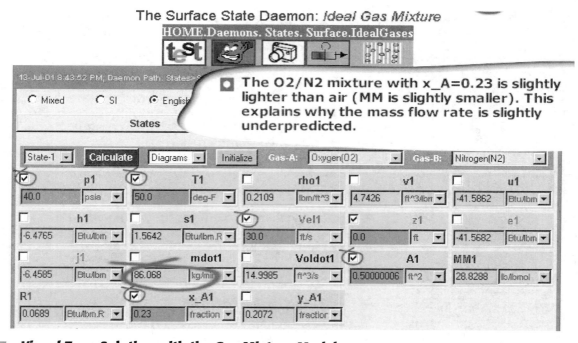

Visual Tour: Solution with the Gas Mixture Model

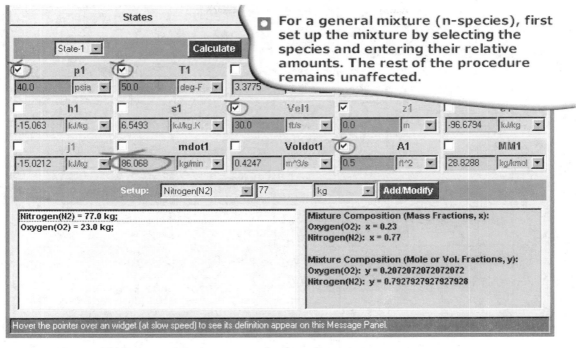

> For a general mixture (n-species), first set up the mixture by selecting the species and entering their relative amounts. The rest of the procedure remains unaffected.

■ *Visual Tour*: **Solution with the General Gas Mixture**

● Let us go over the following moist air problem.

Air is flowing through a duct with a cross-section area of 0.2 m^2 at a dry-bulb temperature of 20 deg-C and relative humidity of 20%. If the pressure inside the duct is 100 kPa and the volume flow rate is 100 m^3/min, determine the flow velocity, dew point temperature and the mass flow rate of dry air and water vapor.

What-if-scenario: How would the answers change if the pressure in the duct is 150 kPa instead?

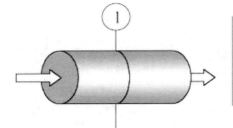

> Solution Procedure: The problem involves finding the state of moist air across a surface area. We choose the States.Surface daemon with Moist Air as the working fluid.

■ *Visual Tour*: **Moist Air, a Special Gas Mixture**

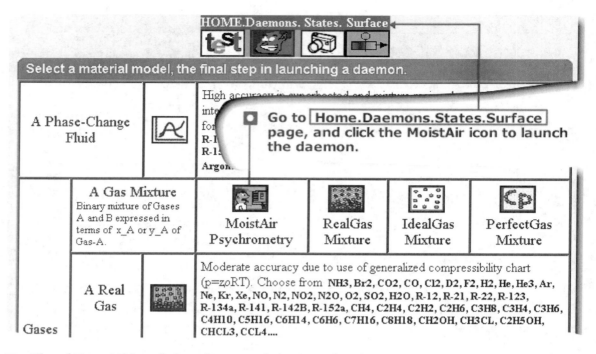

Visual Tour: Moist Air (Psychrometry) State Evaluation

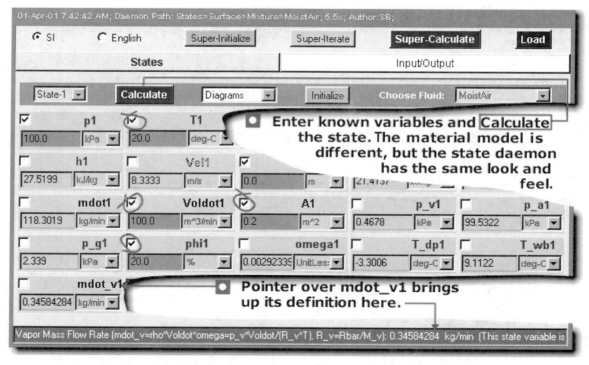

Visual Tour: New Working Fluid Model, but the Same Interface

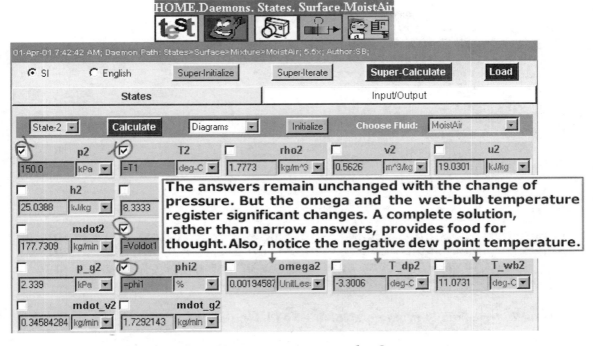

■ *Visual Tour*: What if the Total Pressure Is No1 100 kPa?

● Solid/Liquid Model:

In addition to the specific heats, the density (and, hence, specific volume) is assumed constant and is treated as a material property. If there is no possibility of a phase change in a problem involving a liquid or a solid, this model, like the perfect gas model for gases, serves as an excellent choice, especially in a manual solution.

■ *Visual Tour*: Modeling Pure Solid or Liquid

Example: Compare the total amount of disorder (total entropy) in a container of volume 1 m³ filled with (a) copper, (b) silver, (c) wood, (d) water, or (e) oil. The temperature is maintained at 30 deg C.

How do the variables, p, T, Vel, and z affect the answers?

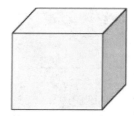

☑ **Solution Procedure:** Simplify the problem as a volume state problem. Select the solid/liquid model to launch the daemon. Enter the known variables and calculate entropy. Multiply by the mass of the system to obtain the total amount of disorder.

■ *Visual Tour*: **A Simple Problem Involving Solid/Liquid Model**

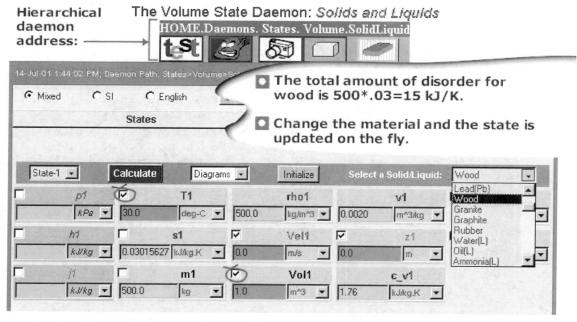

■ *Visual Tour*: **Use the Map to Launch the S/L Volume Daemon**

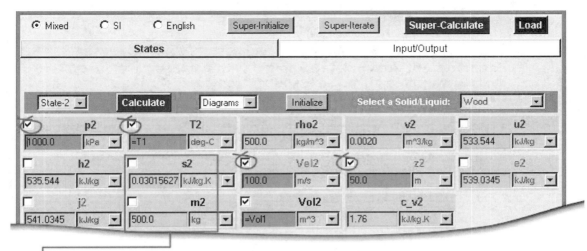

□ State-2, calculated with different values of p, Vel, and z, produces the same entropy and mass. The amount of disorder for the solid/liquid model can be shown to depend only on temperature and specific heat.

■ *Visual Tour*: State-2 Is Derived From State-1

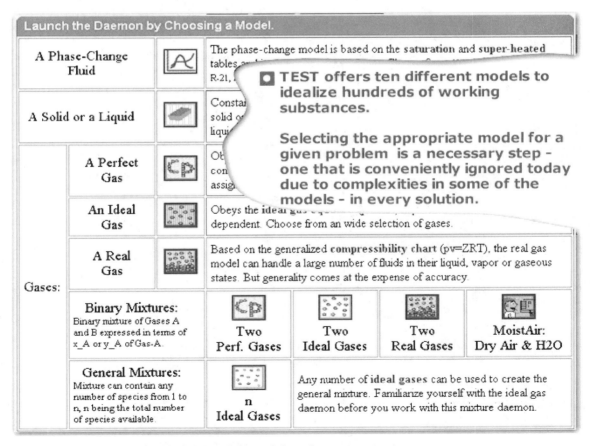

■ *Visual Tour*: Selecting a Material Model Is the Last Step

● The System Daemons:

The next few slides give a bird's eye view of the system daemons. For a complete example with a particular type of system daemons, browse a specific topic listed on the Slide Show home page.

TEST divides all systems into two types: generic systems covered in the first half of most textbooks, and specific systems that address special topics such as power plants, refrigerators, HVAC, combustions, etc.

■ *Visual Tour*: **A World of System Daemons**

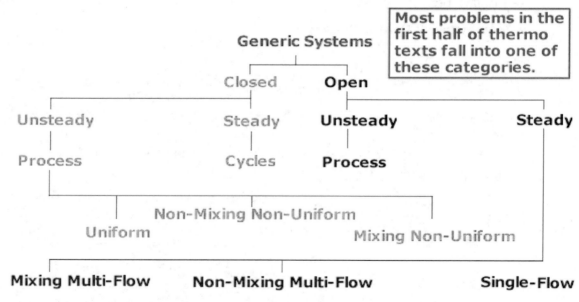

■ *Visual Tour*: **Classification of Generic Systems**

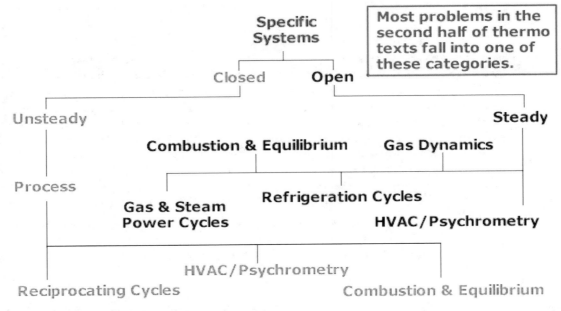

> Most problems in the second half of thermo texts fall into one of these categories.

■ *Visual Tour*: **Classification of Specific Systems**

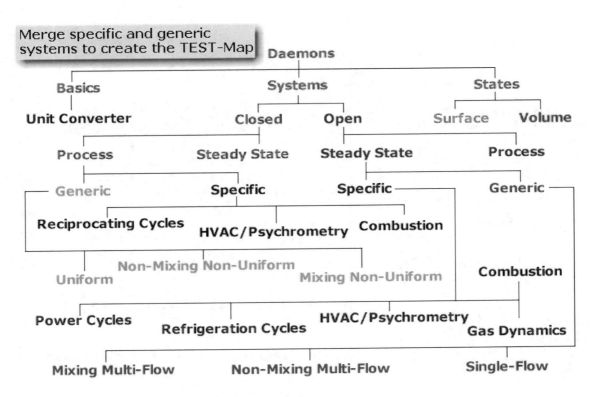

■ *Visual Tour*: **The Entire TEST Daemon Structure**

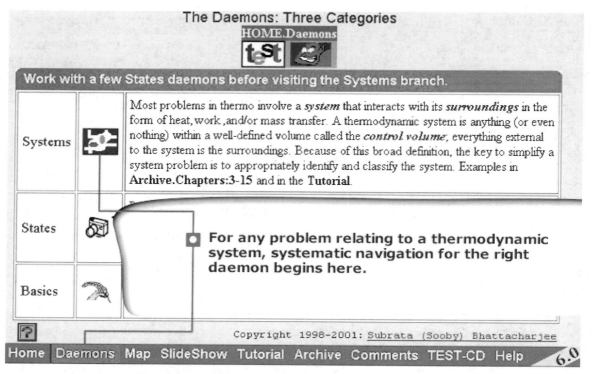

■ *Visual Tour*: **Systematic Simplification**

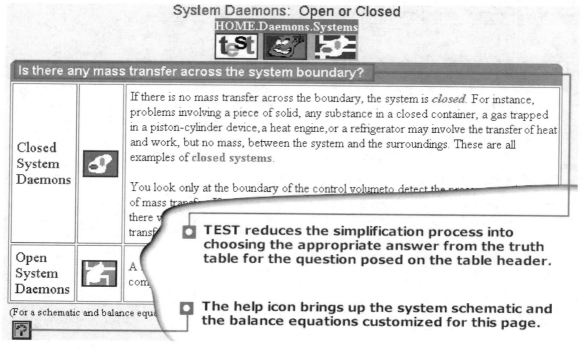

■ *Visual Tour*: **Simplify the System**

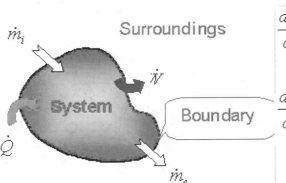

$$\frac{dm}{dt} = \sum_i \dot{m}_i - \sum_e \dot{m}_e$$

$$\frac{dE}{dt} = \sum_i \dot{m}_i e_i - \sum_e \dot{m}_e e_e + \dot{Q} - \dot{W}$$

$$e \equiv u + \frac{V^2}{2} + gz$$

$$\frac{dS}{dt} = \sum_i \dot{m}_i s_i - \sum_e \dot{m}_e s_e + \frac{\dot{Q}}{T_B} + \dot{S}_{gen}$$

- As you simplify the system, the system sketch and the equations adjust to the new specifications.

■ *Visual Tour*: **A Click on the Help Button Brings Up the System and Its Balance Equations**

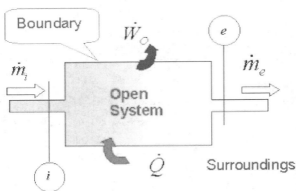

Mass : $\quad 0 = \dot{m}_i - \dot{m}_e$

Energy : $\quad 0 = \dot{m}_i j_i - \dot{m}_e j_e + \dot{Q} - \dot{W}_O$

$$j \equiv h + \frac{V^2}{2} + gz$$

Entropy : $\quad 0 = \dot{m}_i s_i - \dot{m}_e s_e + \frac{\dot{Q}}{T_B} + \dot{S}_{gen}$

- For instance, while solving an open, steady, single-flow problem, the evolution of the system schematic and the governing equations provides insights into the theory behind the System daemons.

■ *Visual Tour*: **Simplification of Sketch and Balance Equations**

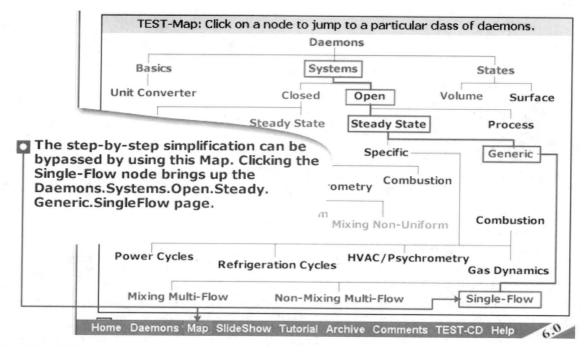

The step-by-step simplification can be bypassed by using this Map. Clicking the Single-Flow node brings up the Daemons.Systems.Open.Steady. Generic.SingleFlow page.

■ **Visual Tour**: TEST-Map—A Shortcut for Simplification

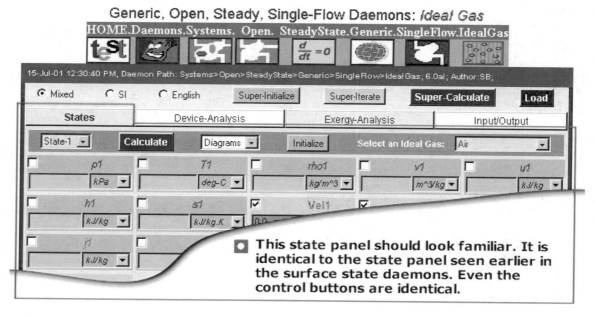

This state panel should look familiar. It is identical to the state panel seen earlier in the surface state daemons. Even the control buttons are identical.

Two new tabs, Device-Analysis and Exergy-Analysis, appear in this daemon.

■ **Visual Tour o**: All System Daemons Have the Familiar States Panel as the Base

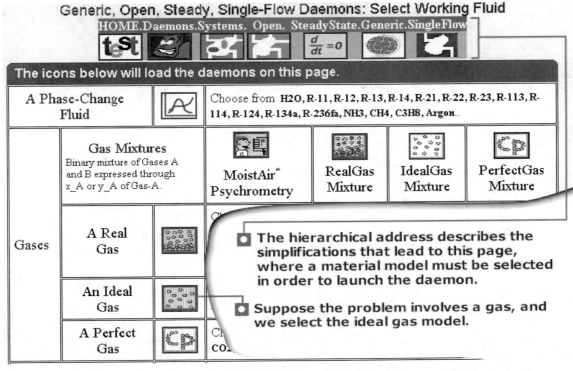

Visual Tour: Last Step to Launch a System Daemon—Decide on a Working Fluid

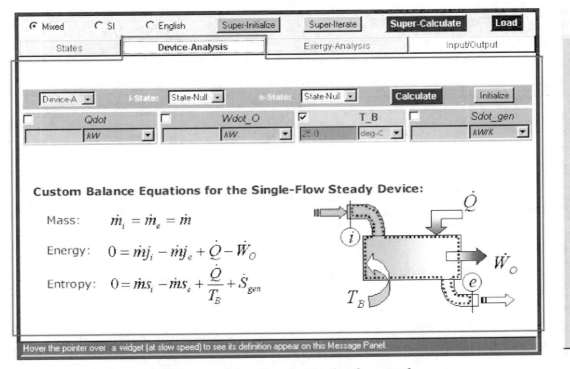

Visual Tour: The Analysis Panel for Steady Single-Flow Devices

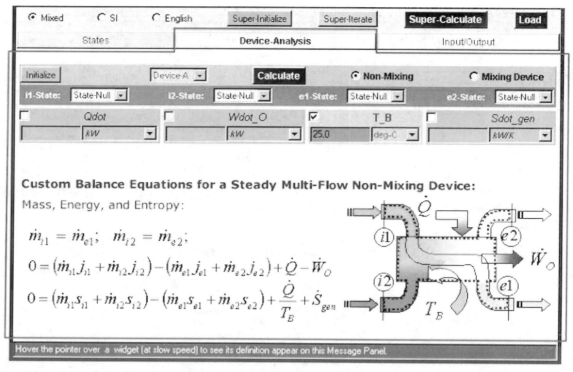

■ *Visual Tour*: **Multi-Flow Non-Mixing Device**

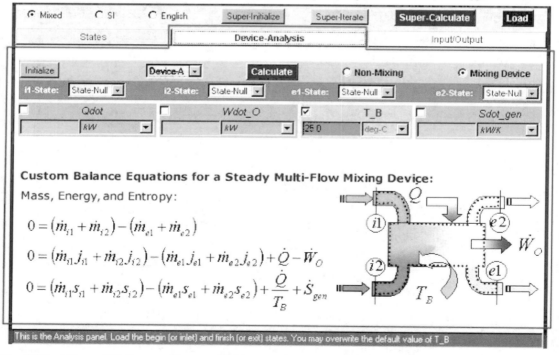

■ *Visual Tour*: **Mixing and Non-Mixing Devices Can Be Toggled**

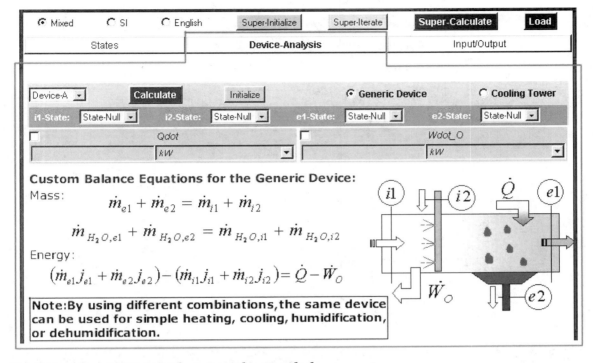

Visual Tour: HVAC Devices Are Also Toggled

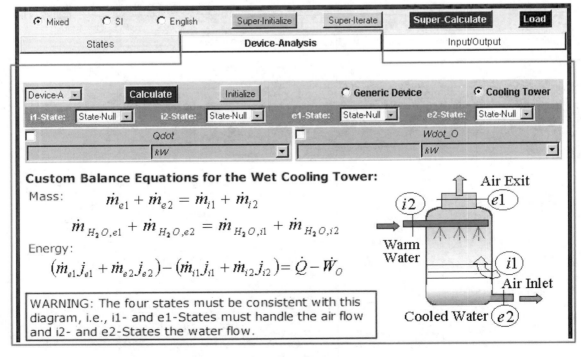

Visual Tour: Device-Specific Analysis—Cooling Tower

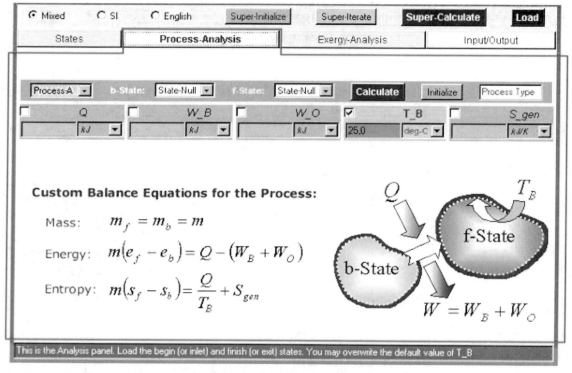

Visual Tour: Analysis Panel for a Closed Process (Uniform System)

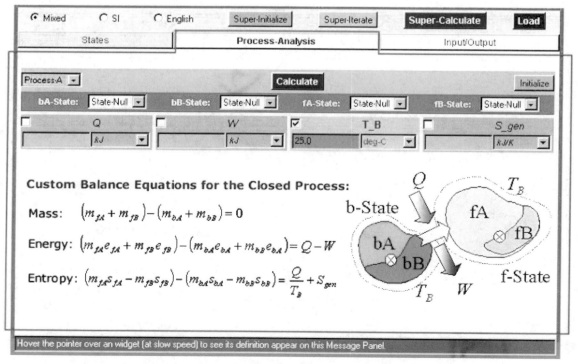

Visual Tour: Analysis Panel for a Closed Process (Non-Uniform System)

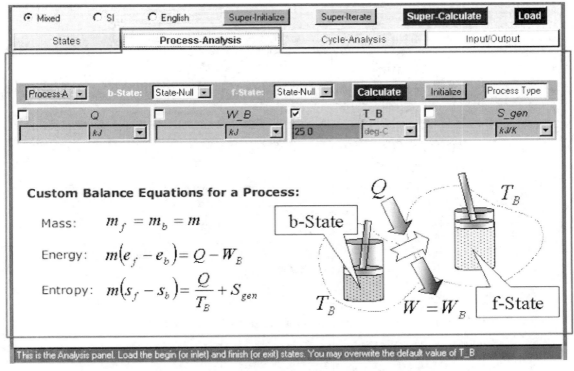

Visual Tour: Analysis Panel for a Process in a Reciprocating Cycle

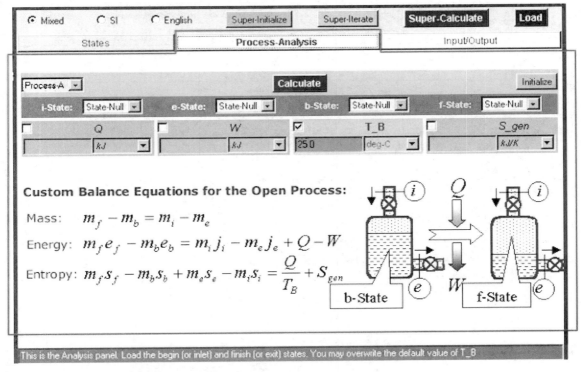

Visual Tour: Analysis Panel for an Open Process

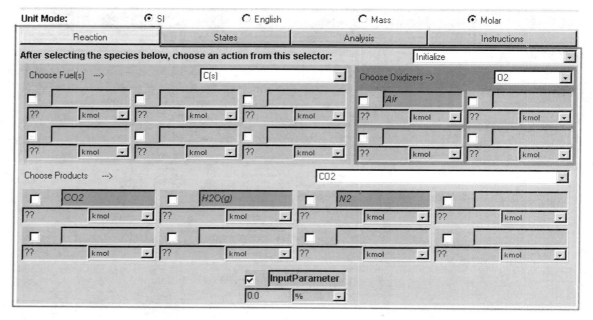

■ *Visual Tour*: **The Combustion Daemon Uses a Special Reaction Panel**

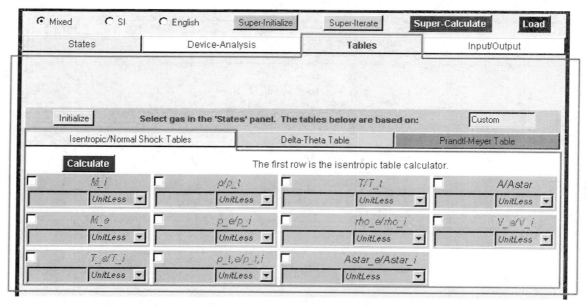

■ *Visual Tour*: **The Gas Dynamics Daemon Offers Many Tables**

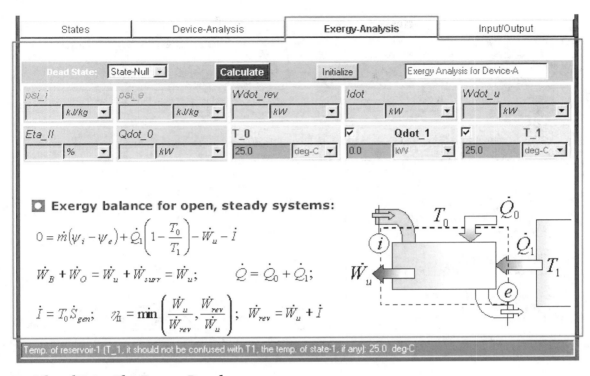

Visual Tour: The Exergy Panel

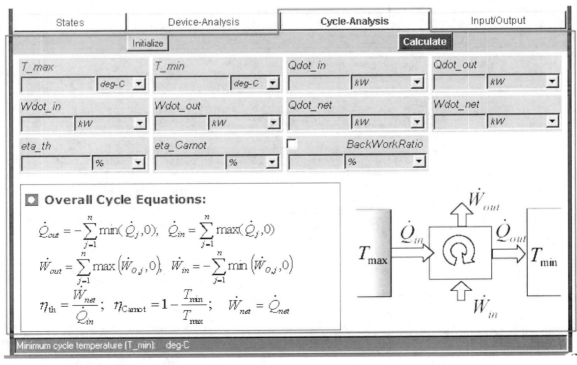

Visual Tour: The Cycle Panel

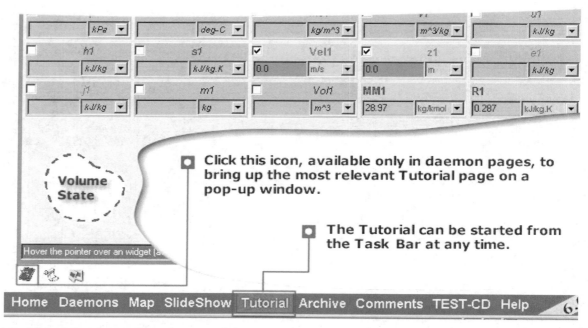

■ ***Visual Tour*: Starting The Tutorial**

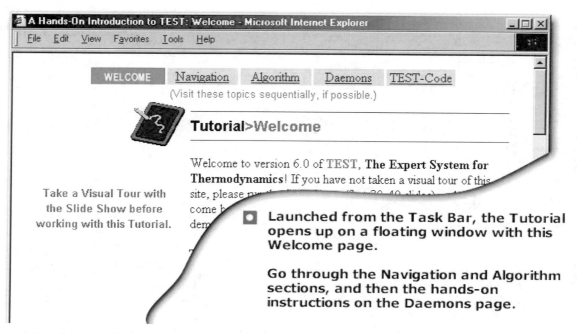

■ ***Visual Tour*: Five Sections of the Tutorial**

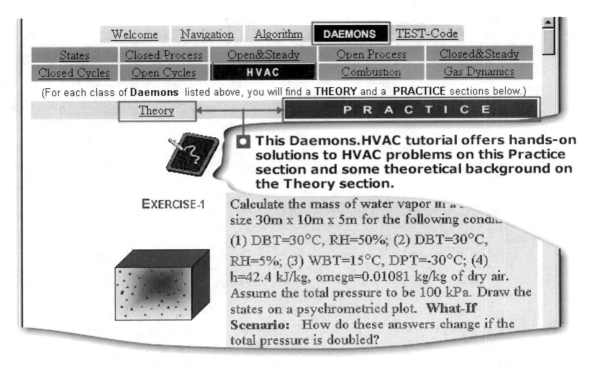

Visual Tour: The Tutorial Icon under a Daemon Pops Up the Practice Page

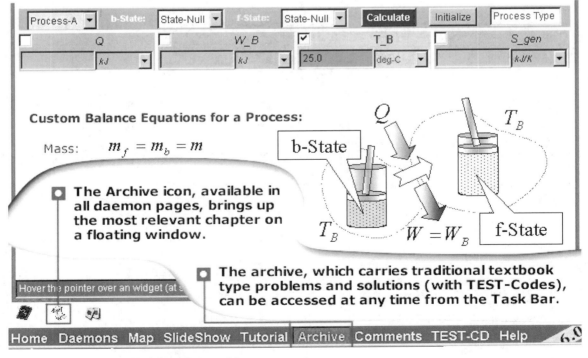

Visual Tour: Launching the Archive

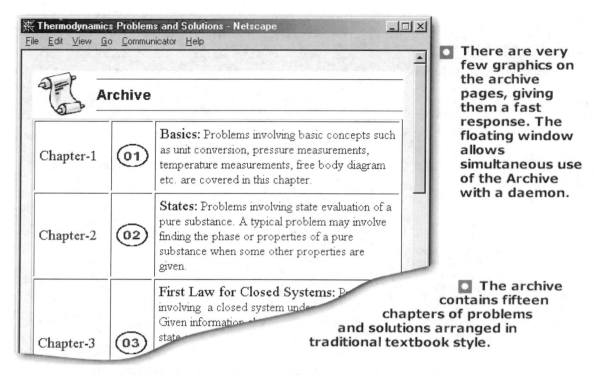

There are very few graphics on the archive pages, giving them a fast response. The floating window allows simultaneous use of the Archive with a daemon.

The archive contains fifteen chapters of problems and solutions arranged in traditional textbook style.

■ *Visual Tour*: The Archive on a Pop-Up Window

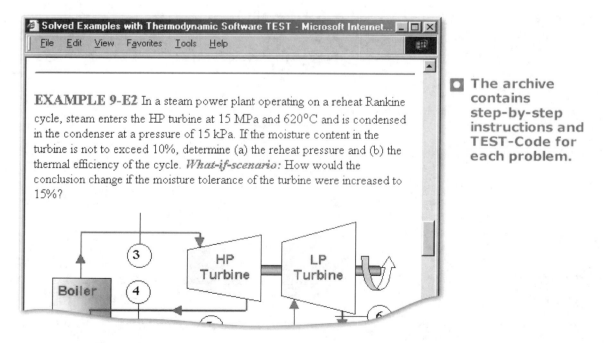

The archive contains step-by-step instructions and TEST-Code for each problem.

■ *Visual Tour*: Solved problems in the archive

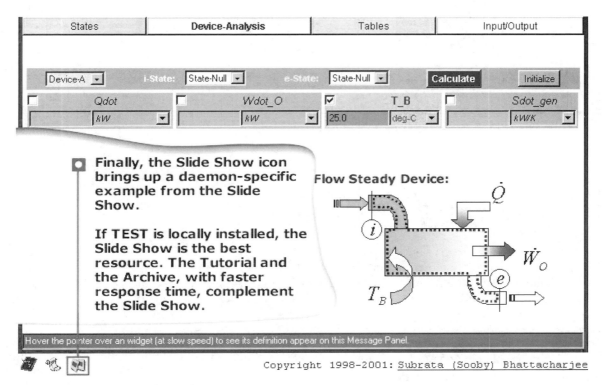

Visual Tour: The Slide Show Icon

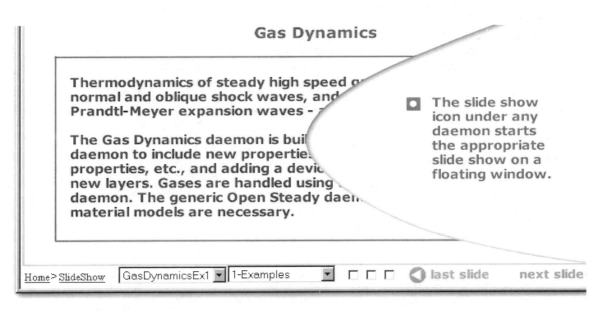

Visual Tour: The Slide Show Icon

States and Properties

TEST extends the concept of a thermodynamic state into an extended state that includes, in addition to the thermodynamic properties (p,T,u,v,h,s, etc.), material (M,R, etc.), extrinsic (e,j,V (Vel), z, etc.), and system (V (Vol), m, etc.) variables. All these variables are color coded consistently throughout TEST for easy identification.

To emphasize the distinction between a control volume and control surface, the State daemons come in two flavors, **Volume State** and **Surface State**, with the system variables mass (m) and volume (V or Vol) replaced with mass flow rate ($\dot{m}$ or mdot) and volume flow rate ($\dot{V}$ or Voldot), respectively, in the latter. For evaluation of the core thermodynamic state, either daemon can be used. It should be emphasized that any combination of independent variables can be used to find a state; pressure, temperature, or even the phase composition, need not be guessed in order to evaluate a state.

There are four models of working substances—Solid/Liquid, Gases, Gas Mixtures, and Phase-Change Fluids—that are available in TEST. Gas models are subdivided into Perfect (constant cp), Ideal, and Real Gas models. The Gas Mixture model allows binary mixtures of perfect, ideal, or real gases, and a general mixture of any number of ideal gases. The composition, however, is assumed frozen once the mixture is specified. Work is currently underway to allow equilibrium composition in a mixture. In a problem-solving session—be it a state evaluation or a system problem—the last step, always, is the selection of a suitable model for the working substance. In selecting a model for H_2O, for instance, a student discovers that H_2O appears under several models—Solid/Liquid, Perfect Gas, Ideal Gas, Real Gas, and Phase-Change models. The ease with which the model can be switched makes TEST a better alternative to the traditional steam table.

The accuracy of the properties calculated is reasonably high (within 1%, except for the Real Gas model). To check the consistency of calculations, use different combinations of input (simply uncheck a known variable and click the check box of a calculated variable) in an evaluated state.

The following examples cover a few selected material models. Note that in TEST the procedure for a state evaluation remains identical whether it is calculations involving a steam table or a psychrometric table. More examples of state evaluation can be found in the Tutorial (specifically written for students) and the Archive section accessible from the Task Bar of the TEST home page.

Example: A piston cylinder device contains a 50–50 (by volume) mixture of N_2 and H_2 at 100 kPa and 30 deg C. The gas mixture is compressed isentropically (s=constant) to half its original volume. Determine the final pressure and temperature.

How would the answers change if the gas is pure H_2 or pure N_2?

🔲 Solution Procedure: Although this problem should be treated as a system problem, because there is no heat or work involved, we will use the States.Volume daemon to evaluate the initial and final states. To model the working fluid we choose the ideal gas mixture model.

■ *States & Properties Ex. #1*: **Problem Description**

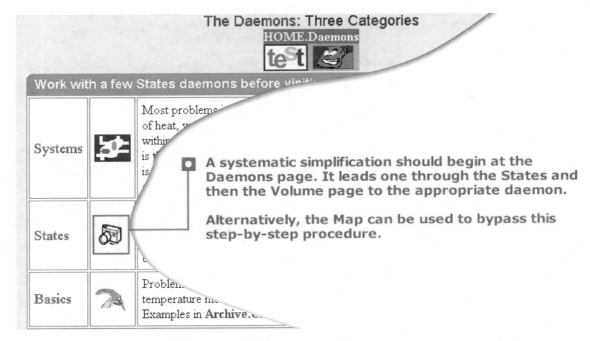

A systematic simplification should begin at the Daemons page. It leads one through the States and then the Volume page to the appropriate daemon.

Alternatively, the Map can be used to bypass this step-by-step procedure.

■ *States & Properties Ex. #1*: **Systematic daemon launch.**

A piston cylinder device contains... Click on the daemon link on the Task Bar to bring up the daemon page. Follow the directions in the simplification table.

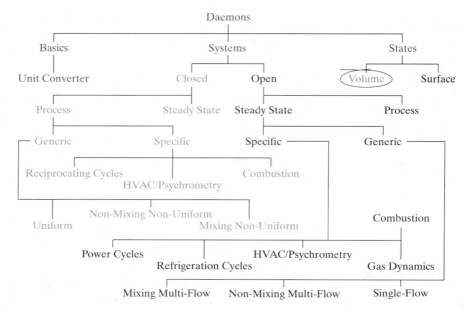

A piston cylinder device contains... This is a state evaluation problem involving a closed volume. The appropriate daemon can be found in the following page: Daemons.States.Volume.

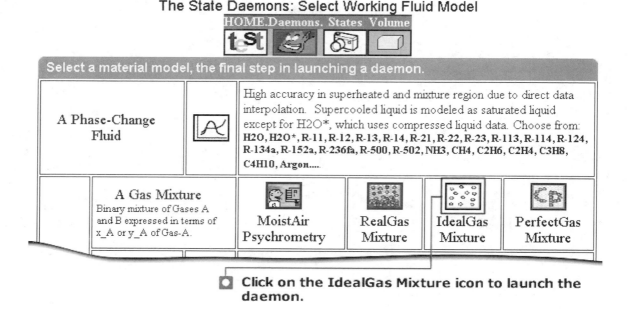

■ *States & Properties Ex. #1:* **Idealization of Working Material**

A piston cylinder device contains a 50–50 mixture of... We select the ideal gas mixture to model the working fluid. The perfect gas mixture model, although less accurate, will also work.

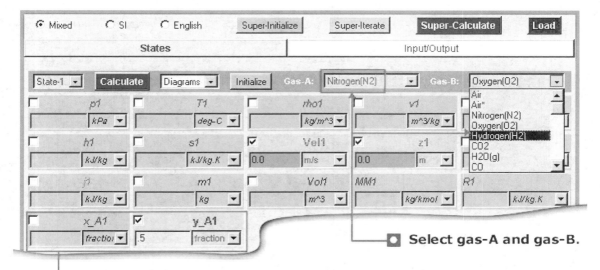

Select gas-A and gas-B.

For any mixture, the composition must be specified in terms of mass fraction, x_A, or mole fraction, y_A, of gas-A. Here we specify the mole fraction which is same as the volume fraction for gas mixtures.

■ *States & Properties Ex. #1*: Compose the Mixture

A piston cylinder device contains a 50–50 mixture of. . . Select N2 as gas-A and H2 as gas-B, and specify y_A, the mole fraction of gas-A.

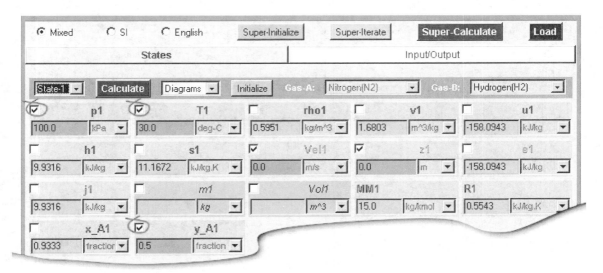

Enter the known variables and Calculate state-1. The default values of Vel and z are left unchanged. Note that the mixture molecular mass (MM) and the gas constant (R) are calculated as part of the state.

■ *States & Properties Ex. #1*: Evaluate State-1

A piston cylinder device contains a 50–50 mixture of. . . Without x_A (mass fraction of gas-A) or y_A (mole or volume fraction of gas-A) a mixture state cannot be evaluated.

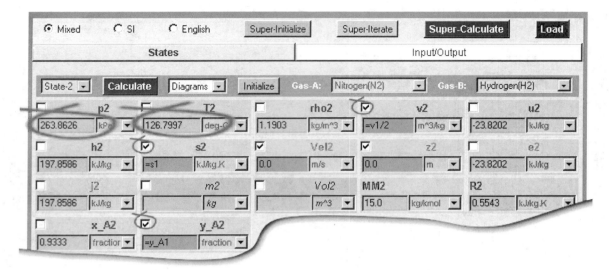

□ **Use of equations instead of absolute values in state-2 makes it easy to perform parametric studies.**

■ *States & Properties Ex. #1*: Evaluate State-2

A piston cylinder device contains a 50–50 mixture of... Enter y_A2, v2(= v1/2), and s2, and Calculate the desired unknowns along with the rest of the state.

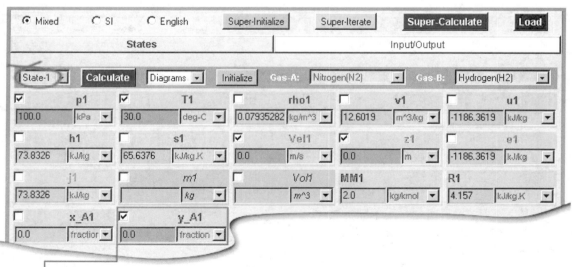

□ **To convert the mixture into pure hydrogen, make gas-A disappear by assigning y_A1=0.0. Calculate state-1 and Super-Calculate to update all others.**

■ *States & Properties Ex. #1*: Change a Parameter–Mixture Composition

A piston cylinder device contains a 50–50 mixture of... By changing x_A or y_A to 0 or the mixture can be converted into pure nitrogen or pure hydrogen.

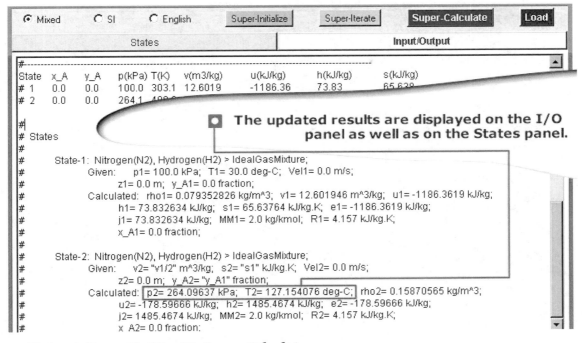

■ *States & Properties Ex. #1*: **Super-Calculate**

A piston cylinder device contains a 50–50 mixture of... The Super-Calculate operation automatically takes you to the I/O panel, where the detailed report is displayed.

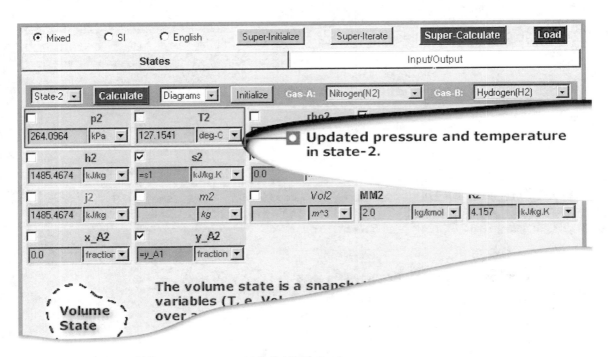

■ *States & Properties Ex. #1*: **An Updated State-2**

A piston cylinder device contains a 50–50 mixture of... Go back to the States panel and select State-2. The Super-Calculate operation has updated the state. Note that the change in composition does not affect the answers significantly.

```
######################################################################
# To regenerate this solution, copy the following TEST-Code onto the I/O panel of the
# ..States.Volume.IdealGasMixture daemon.
# and click the Load and Super-Calculate buttons.
#-------------------------------------Start of TEST-Code-------------------------------------

States {
State-1: Nitrogen(N2), Hydrogen(H2);
Given:        { p1= 100.0 kPa;  T1= 30.0 deg-C; Vel1= 0.0 m/s; z1= 0.0 m; y_A1= 0.5
fraction; }

State-2: Nitrogen(N2), Hydrogen(H2);
Given:        { v2= "v1/2" m^3/kg;  s2= "s1" kJ/kg.K;  Vel2= 0.0 m/s; z2= 0.0 m; y_A2=
"y_A1" fraction; }
}

#-------------------------------------End of TEST-Code-------------------------------------
--------
```

■ *States & Properties Ex. #1: TEST-Code to Regenerate This Solution*

Example: Carbon dioxide at 5 atm and 150 deg C is to be supplied at a mass flow rate of 0.5 kg/min flowing through a tube with an area of cross section of 2 cm^2. Determine the velocity of the flow.

What-If Scenario: How would the answers change if the working fluid is helium instead?

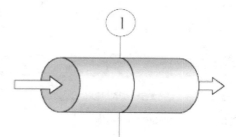

☐ Solution Procedure: Velocity is an extrinsic property of a state. The state across the area of cross section can be assumed uniform. Launch the States.Surface daemon, enter the known variables, and Calculate the velocity.

■ *States & Properties Ex. #2: Problem Description*

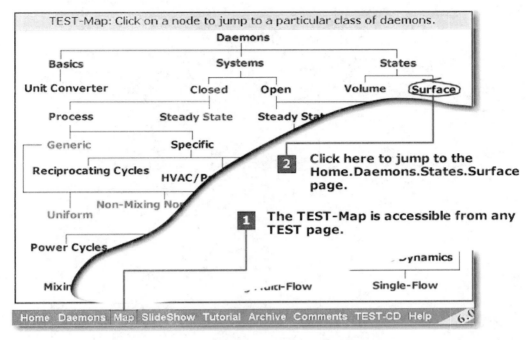

States & Properties Ex. #2: Shortcut for the Experienced User

Carbon dioxide at 5 atm and 150 deg-C is to be supplied... The Daemons.States.Surface page can be directly reached through the TEST-Map.

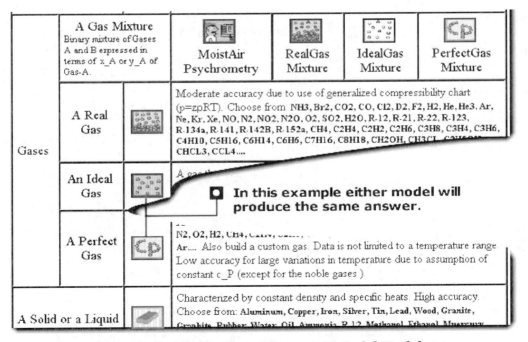

States & Properties Ex. #2: Idealization—Choose a Material Model

Carbon dioxide at 5 atm and 150 deg-C is to be supplied... If there is no temperature variation in a problem or if the gas is an inert gas, the perfect gas model performs as accurately as the ideal gas model. We chose the perfect gas model in this problem.

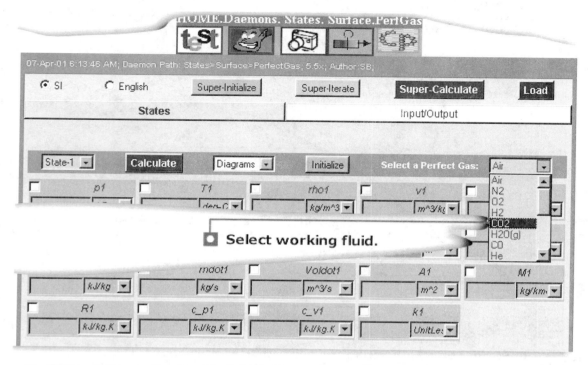

States & Properties Ex. #2: Select CO2 as the Working Fluid

Carbon dioxide at 5 atm and 150 deg-C is to be supplied... Select CO2 from the species selector.

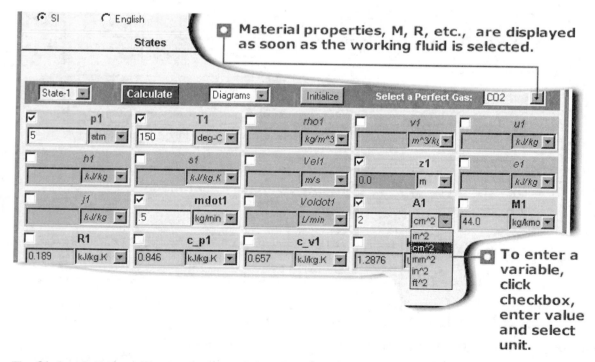

States & Properties Ex. #2: Enter Known Variables

Carbon dioxide at 5 atm and 150 deg-C is to be supplied... Default values of zero are assigned to variables Vel and z. Because Vell is the desired unknown, uncheck its box to overwrite the default value.

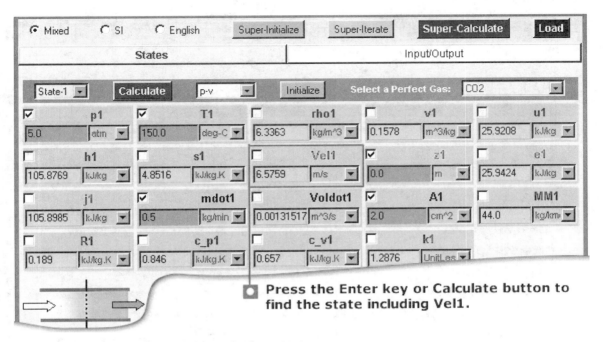

Press the Enter key or Calculate button to find the state including Vel1.

■ *States & Properties Ex. #2*: **Calculate State-1**

Carbon-dioxide at 5 atm and 150 deg-C is to be supplied. . .

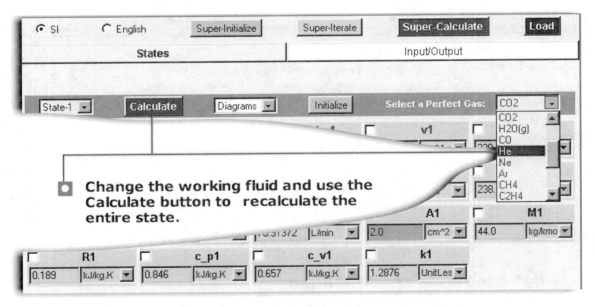

Change the working fluid and use the Calculate button to recalculate the entire state.

■ *States & Properties Ex. #2*: **What-If Study—Change Working Fluid**

How would the answers change if the working fluid were helium instead?

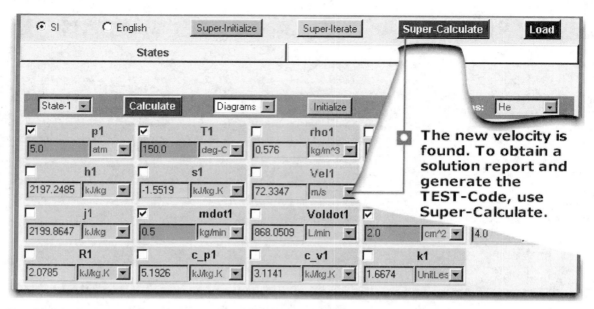

■ *States & Properties Ex. #2:* **What-If Study—Calculate New State**

How would the answers change if the working fluid were helium instead?

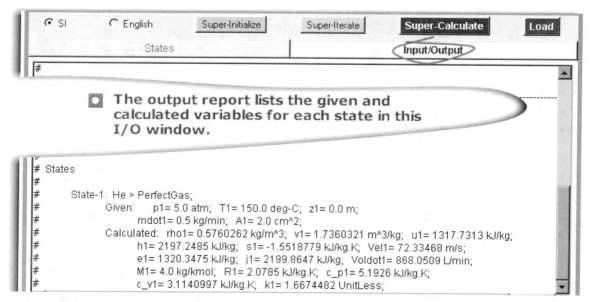

■ *States & Properties Ex. #2:* **Solution Report**

Carbon dioxide at 5 atm and 150 deg-C is to be supplied... The Super-Calculate operation automatically changes the focus to the I/O panel, where the solution report and the TEST-Codes are generated.

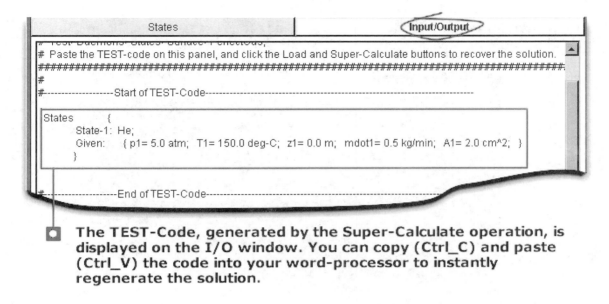

The **TEST-Code**, generated by the Super-Calculate operation, is displayed on the I/O window. You can copy (Ctrl_C) and paste (Ctrl_V) the code into your word-processor to instantly regenerate the solution.

■ *States & Properties Ex. #2:* **TEST-Code**

Carbon dioxide at 5 atm and 150 deg-C is to be supplied. Saving the TEST-Code allows one to jump-start the solution at a later session.

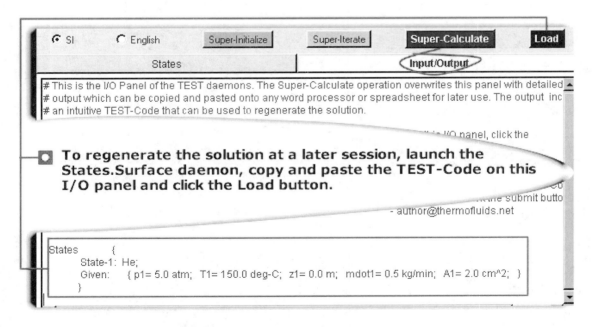

To regenerate the solution at a later session, launch the States.Surface daemon, copy and paste the TEST-Code on this I/O panel and click the Load button.

■ *States & Properties Ex. #2:* **Loading TEST-Code**

Carbon dioxide at 5 atm and 150 deg-C is to be supplied. . . To load TEST-Code, start the same daemon that created the code—States.Surface.Perfect Gas. in this case. Paste the codes on this I/O panel. Note that you can type in the code, too.

◻ **A pop-up window shows up with error messages, if any. Close this floating window.** ─────────

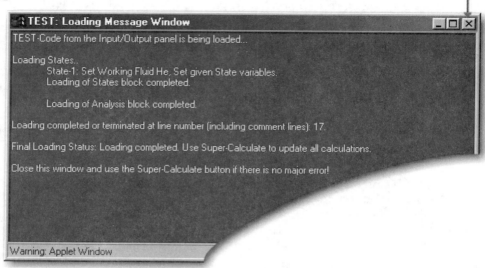

◼ *States & Properties Ex. #2*: **Loading TEST-Code**

Carbon dioxide at 5 atm and 150 deg-C is to be supplied... If the code is pasted (not typed), there should not be any loading error.

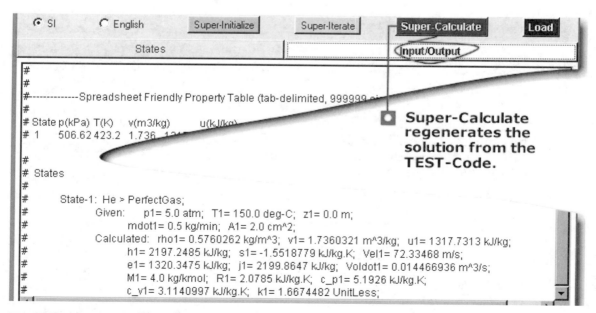

Super-Calculate regenerates the solution from the TEST-Code.

◼ *States & Properties Ex. #2*: **Super-Calculate**

Carbon dioxide at 5 atm and 150 deg-C is to be supplied...

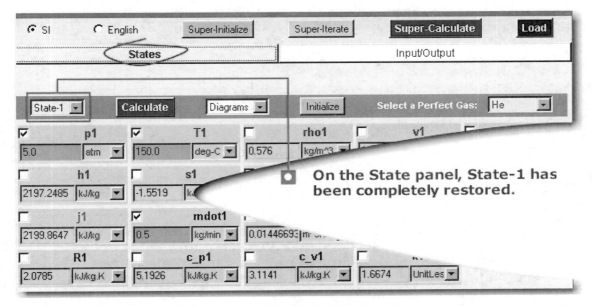

States & Properties Ex. #2: Visual Solution Regenerated

Carbon dioxide at 5 atm and 150 deg-C is to be supplied... All States, processes, or devices, are restored after the TEST-Code is loaded.

```
####################################################################
# To regenerate this solution, copy the following TEST-Code onto the I/O panel of the
# ..States.Surface.PerfectGas daemon
# and click the Load and Super-Calculate buttons.
#-----------------------------------Start of TEST-Code----------------------------------------


States {
State-1: CO2;
   Given:      { p1 = 5.0 atm;  T1 = 150.0 deg-C;  z1 = 0.0 m;  mdot1 = 0.5 kg/min;
               A1 = 2.0 cm^2; }
}


#-----------------------------------End of TEST-Code-----------------------------------------
--------
```

State Evaluation Ex. #2: TEST-Code to Regenerate This Solution

Example: A 10 gallon tank is filled with 15 pounds of propane. If the temperature is 30 deg-C, determine the pressure in atm. If the tank can withstand a pressure of 100 atm, determine the critical temperature at which the tank will fail.

What-If Scenario: How would your answer change if the tank held 20 lbs of propane instead? Re-solve the problem with ethane (C_2H_6) as the working fluid.

> **■ Solution Procedure:** The state across the volume of the tank is assumed uniform. Launch the States.Volume daemon, select Real Gas as the material model, choose propane and evaluate the volume state. Note that Phase-Change model will produce more accurate answers than the generalized compressibility data.

■ *States & Properties Ex. #3:* **Problem Description**

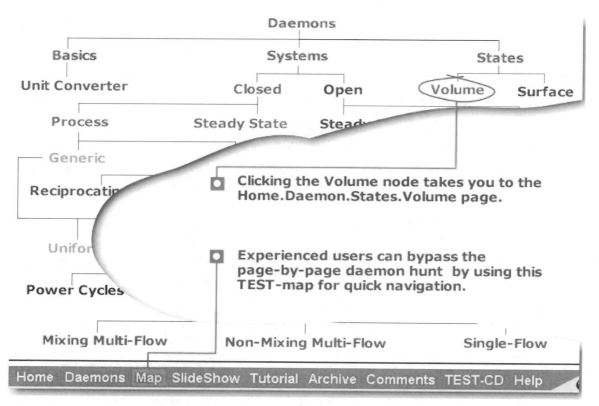

Daemons

Basics   Systems   States

Unit Converter   Closed   Open   Volume   Surface

Process   Steady State   Stead

Generic

Reciprocatin

> ■ Clicking the Volume node takes you to the Home.Daemon.States.Volume page.

Unifor

> ■ Experienced users can bypass the page-by-page daemon hunt by using this TEST-map for quick navigation.

Power Cycles

Mixing Multi-Flow   Non-Mixing Multi-Flow   Single-Flow

Home   Daemons   Map   SlideShow   Tutorial   Archive   Comments   TEST-CD   Help

■ *State Evaluation Ex. #2:* **Direct Navigation Using the TEST-Map**
A 10-gallon tank is filled with 15 lb of propane...

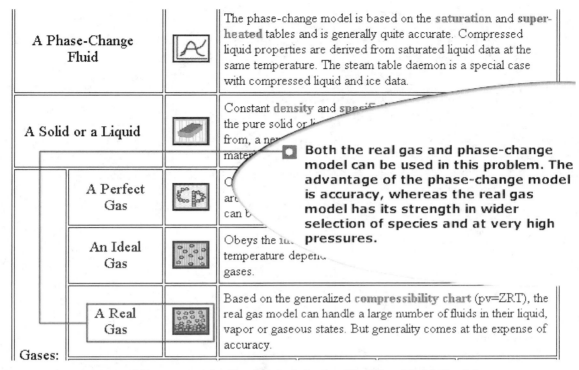

A Phase-Change Fluid		The phase-change model is based on the **saturation** and **super-heated** tables and is generally quite accurate. Compressed liquid properties are derived from saturated liquid data at the same temperature. The steam table daemon is a special case with compressed liquid and ice data.
A Solid or a Liquid		Constant **density** and spe... the pure solid or... from, a ne... mate...
	A Perfect Gas	C... ar... can b...
	An Ideal Gas	Obeys the i... temperature depen... gases.
Gases:	A Real Gas	Based on the generalized **compressibility chart** (pv=ZRT), the real gas model can handle a large number of fluids in their liquid, vapor or gaseous states. But generality comes at the expense of accuracy.

Both the real gas and phase-change model can be used in this problem. The advantage of the phase-change model is accuracy, whereas the real gas model has its strength in wider selection of species and at very high pressures.

■ **States & Properties Ex. #3: Idealization - Select a Working Fluid Model**

A 10 gallon tank is filled with 15 lb of propane...

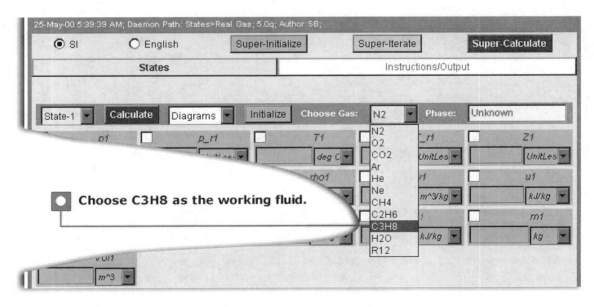

Choose C3H8 as the working fluid.

■ **States & Properties Ex. #3: Select the Working Fluid**

A 10-gallon tank is filled with 15 lb of propane...

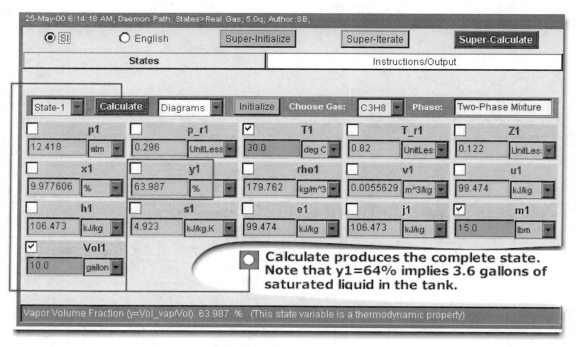

■ **States & Properties Ex. #3: Evaluate State-1**

A 10-gallon tank is filled with 15 lb of propane...

■ **States & Properties Ex. #3: Calculate State-1**

A 10-gallon tank... If the temperature is 30 deg-C, determine the pressure in atm. As you click the Calculate button, the pressure is evaluated as 12.4 atm.

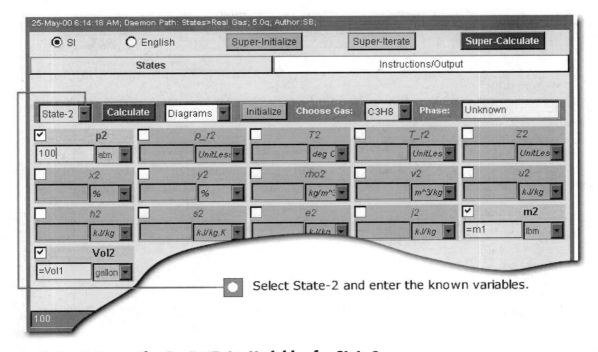

Select State-2 and enter the known variables.

■ *States & Properties Ex. #3:* **Enter Variables for State 2**

If the tank can withstand a pressure of 100 atm, determine the critical temperature at which the tank will fail.

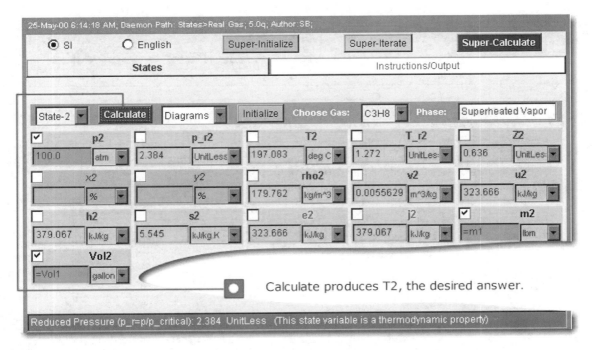

Calculate produces T2, the desired answer.

Reduced Pressure (p_r=p/p_critical): 2.384 UnitLess (This state variable is a thermodynamic property)

■ *States & Properties Ex. #3:* **State-2 Yields the Failure Temperature**

If the tank can withstand a pressure of 100 atm, determine the critical temperature at which the tank will fail. Note that the phase composition is calculated as part of the state.

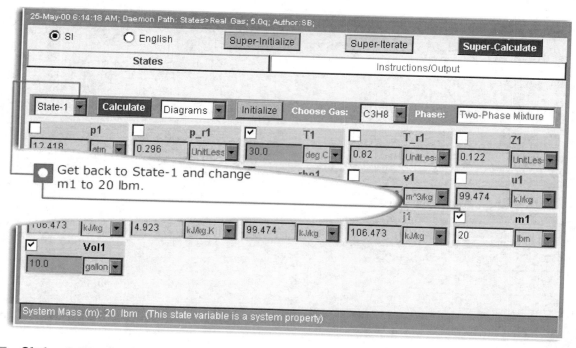

States & Properties Ex. #3: What-If Scenario—Effect of Total Mass

How would your answer change if the tank held 20 lb of propane instead?

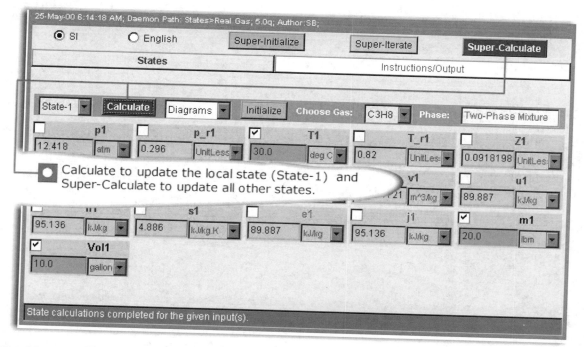

States & Properties Ex. #3: Super-Calculate

How would your answer change if the tank held 20 lb of propane instead?

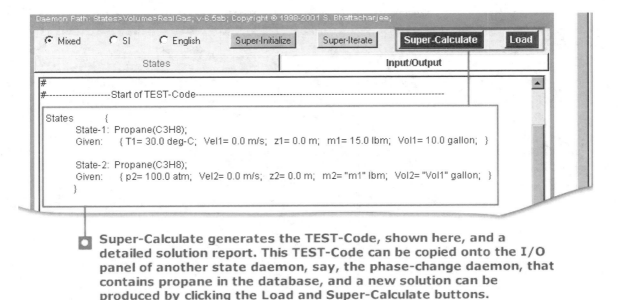

☐ **Super-Calculate generates the TEST-Code, shown here, and a detailed solution report. This TEST-Code can be copied onto the I/O panel of another state daemon, say, the phase-change daemon, that contains propane in the database, and a new solution can be produced by clicking the Load and Super-Calculate buttons.**

■ *States & Properties Ex. #3*: **Output from Super-Calculate Operation**

A 10 gallon tank is filled with 15 pounds of propane... The new failure temperature is only deg-C (435 K). It can be found on this I/O panel or in the States panel.

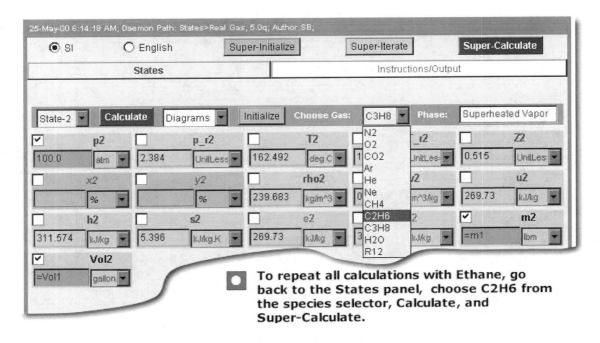

☐ **To repeat all calculations with Ethane, go back to the States panel, choose C2H6 from the species selector, Calculate, and Super-Calculate.**

■ *States & Properties Ex. #3*: **Change the Working Fluid to Ethane**

Resolve the problem with ethane (C2H6) as the working fluid.

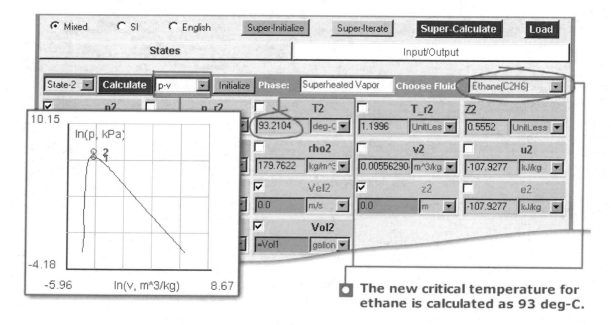

The new critical temperature for ethane is calculated as 93 deg-C.

■ *States & Properties Ex. #3:* All Calculations Updated for C2H6

Resolve the problem with ethane (C2H6) as the working fluid... The new failure temperature is evaluated as 71 deg C.

```
########################################################################
# To regenerate this solution, copy the following TEST-Code onto the I/O panel of the
# ..States.Surface.PerfectGas daemon
# and click the Load and Super-Calculate buttons.
#--------------------Start of TEST-Code----------------------------------------------------------
----

 States {
 State-1: Propane(C3H8);
 Given:      { T1= 30.0 deg-C;  Vel1= 0.0 m/s;  z1= 0.0 m;  m1= 15.0 lbm;  Vol1= 10.0
gallon;  }

 State-2: Propane(C3H8);
 Given:      { p2= 100.0 atm;  Vel2= 0.0 m/s;  z2= 0.0 m;  m2= "m1" lbm;  Vol2= "Vol1"
gallon;  }
 }
```

■ *States & Properties Ex. #3:* TEST-Code to Generate the Visual Solution

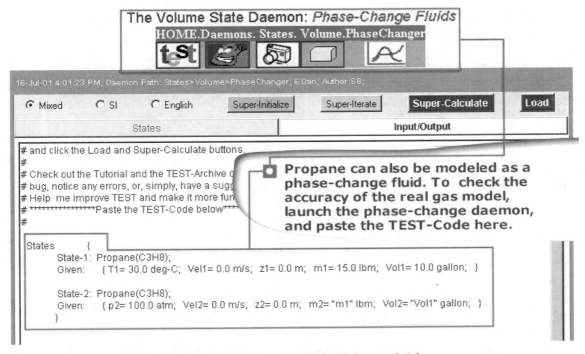

The Volume State Daemon: *Phase-Change Fluids*

HOME.Daemons. States. Volume.PhaseChanger

16-Jul-01 4:01:23 PM; Daemon Path: States>Volume>PhaseChanger; 6.0an; Author:SB;

Propane can also be modeled as a phase-change fluid. To check the accuracy of the real gas model, launch the phase-change daemon, and paste the TEST-Code here.

```
States     {
     State-1: Propane(C3H8);
     Given:    { T1= 30.0 deg-C;  Vel1= 0.0 m/s;  z1= 0.0 m;  m1= 15.0 lbm;  Vol1= 10.0 gallon;  }

     State-2: Propane(C3H8);
     Given:    { p2= 100.0 atm;  Vel2= 0.0 m/s;  z2= 0.0 m;  m2= "m1" lbm;  Vol2= "Vol1" gallon;  }
     }
```

■ *States & Properties Ex. #3*: **Propane as a Phase Change Fluid**

A 10-gallon tank. . . If the temperature is 30 deg-C, determine the pressure in atm. . . Real gas model sacrifices accuracy for the sake of generality. To test that, the TEST-Code can be used to quickly generate the solution with the phase-change model.

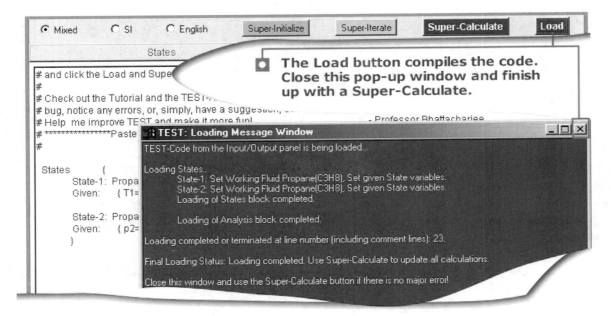

The Load button compiles the code. Close this pop-up window and finish up with a Super-Calculate.

TEST: Loading Message Window

TEST-Code from the Input/Output panel is being loaded...

Loading States..
     State-1: Set Working Fluid Propane(C3H8). Set given State variables.
     State-2: Set Working Fluid Propane(C3H8). Set given State variables.
     Loading of States block completed.

     Loading of Analysis block completed.

Loading completed or terminated at line number (including comment lines): 23.

Final Loading Status: Loading completed. Use Super-Calculate to update all calculations.

Close this window and use the Super-Calculate button if there is no major error!

■ *States & Properties Ex. #3*: **Loading the TEST-Codes**

A 10-gallon tank. . . If the temperature is 30 deg-C, determine the pressure in atm. After copying the TEST-Code, use the Load button.

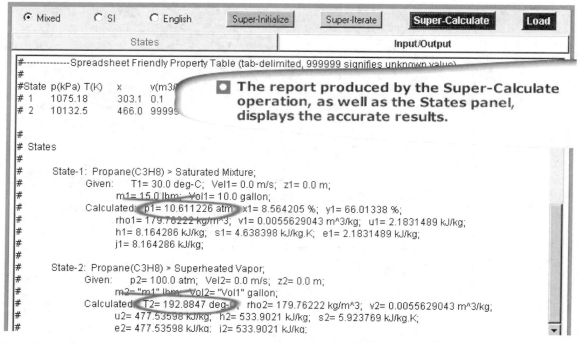

The report produced by the Super-Calculate operation, as well as the States panel, displays the accurate results.

```
#------------Spreadsheet Friendly Property Table (tab-delimited, 999999 signifies unknown value)
#
#State p(kPa) T(K)    x     v(m3/
# 1   1075.18  303.1  0.1
# 2   10132.5  466.0  99999
#
# States
#
#     State-1: Propane(C3H8) > Saturated Mixture;
#         Given:    T1= 30.0 deg-C;  Vel1= 0.0 m/s;  z1= 0.0 m;
#                   m1= 15.0 lbm;  Vol1= 10.0 gallon;
#         Calculated:  p1= 10.611226 atm;  x1= 8.564205 %;  y1= 66.01338 %;
#                   rho1= 179.76222 kg/m^3;  v1= 0.0055629043 m^3/kg;  u1= 2.1831489 kJ/kg;
#                   h1= 8.164286 kJ/kg;  s1= 4.638398 kJ/kg.K;  e1= 2.1831489 kJ/kg;
#                   j1= 8.164286 kJ/kg;
#
#     State-2: Propane(C3H8) > Superheated Vapor;
#         Given:    p2= 100.0 atm;  Vel2= 0.0 m/s;  z2= 0.0 m;
#                   m2= "m1" lbm;  Vol2= "Vol1" gallon;
#         Calculated:  T2= 192.8847 deg-C;  rho2= 179.76222 kg/m^3;  v2= 0.0055629043 m^3/kg;
#                   u2= 477.53598 kJ/kg;  h2= 533.9021 kJ/kg;  s2= 5.923769 kJ/kg.K;
#                   e2= 477.53598 kJ/kg;  i2= 533.9021 kJ/kg;
```

■ *States & Properties Ex. #3*: Super-Calculate the New Answers

A 10-gallon tank. . . If the temperature is 30 deg-C, determine the pressure in atm. After loading the TEST-Code, click Super-Calculate.

Balance Equations

The systematic simplification of the governing balance equations for mass, energy, and entropy is illustrated with an example in this chapter. The example illustrates that to derive the governing equations for a mixing problem, one does not have to be knowledgeable about mixture properties such as the mass or mole fraction. This type of exercise, which can be performed for almost any system problem, can be used to develop the habit of separating the fundamental laws of thermodynamics from the mechanics of property evaluation. Perhaps educators need no reminder that students spend a disproportionate amount of time in the latter, which takes the fun out of the subject for many.

Example: Nitrogen at 500 kPa and 50 deg-C enters an adiabatic mixing chamber with a flow rate of 1 kg/s, while helium enters at 500 kPa, 500 deg-C and 2 kg/s. There is no pressure drop in the chamber and KE and PE can be assumed negligible.

Starting from a generic system show how the balance equation simplifies for this problem.

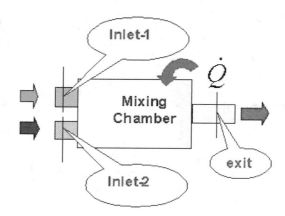

◘ **Solution Procedure: Start with the Daemons.Systems page. Use the help icon to display the system schematic and governing equations. Simplify the problem using the simplification tables, and at each step click on the help icon to monitor the evolving system. The analysis panel of the daemon contains the final form of the balance equations.**

■ *Balance Equations Ex. #1:* **Problem Description**

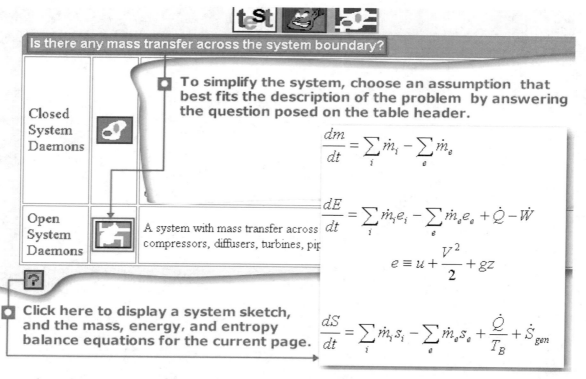

■ *Balance Equations Ex. #1:* **The Most General System**

Nitrogen at 500 kPa and 50 deg-C enters a mixing. . . The Daemons.Systems page offers two choices for simplification of the system.

Open System Daemons: Open Steady or Open Process Problems

HOME.Daemons.Systems. Open

Does the state of the system remain frozen in time?

Page	Icon	Description
Open Steady	$\frac{d}{dt} = 0$	The snapshot of the system taken with a State Camera does not change with time (read the **Tutorial**>**Algorithm** page) when a system is at *steady state*. The total mass, energy, and entropy of the system do not change with time as a result. The bulk of open system problems involving devices such as nozzles, turbines, pumps, compressors, diffusers, condensers, evaporators, etc., belong to this category.
Open Unsteady Instantaneous		
Open Process	$_b\int^f$	If the open system is ~~state, the hallmark of any process~~, it executes ~~b~~ and f-states as found in a closed process, there must be an inlet or exit port, characterized by the i-state or e-state in an open process. Charging a propane cylinder is an example of an open process.

☐ The picture of the mixing chamber does not change with time. Hence, we make the steady-state assumption.

■ *Balance Equations Ex. #1:* **The Systems.Open Page**

Nitrogen at 500 kPa and 50 deg-C enters a mixing… The open system equations can be displayed by clicking the Help button found below the simplification table.

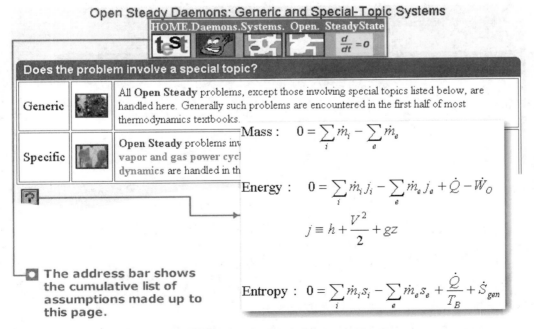

Open Steady Daemons: Generic and Special-Topic Systems

HOME.Daemons.Systems. Open. SteadyState $\frac{d}{dt} = 0$

Does the problem involve a special topic?

Generic		All **Open Steady** problems, except those involving special topics listed below, are handled here. Generally such problems are encountered in the first half of most thermodynamics textbooks.
Specific		**Open Steady** problems inv~~olving~~ vapor and gas power cycl~~es~~ dynamics are handled in th~~e~~

$$\text{Mass}: \quad 0 = \sum_i \dot{m}_i - \sum_e \dot{m}_e$$

$$\text{Energy}: \quad 0 = \sum_i \dot{m}_i j_i - \sum_e \dot{m}_e j_e + \dot{Q} - \dot{W}_O$$

$$j \equiv h + \frac{V^2}{2} + gz$$

$$\text{Entropy}: \quad 0 = \sum_i \dot{m}_i s_i - \sum_e \dot{m}_e s_e + \frac{\dot{Q}}{T_B} + \dot{S}_{gen}$$

☐ The address bar shows the cumulative list of assumptions made up to this page.

■ *Balance Equations Ex. #1:* **The Systems.Open.Steady Page**

Nitrogen at 500 kPa and 50 deg-C enters a mixing. . . The time derivatives disappear as the total mass, energy, and entropy of the system remain constant.

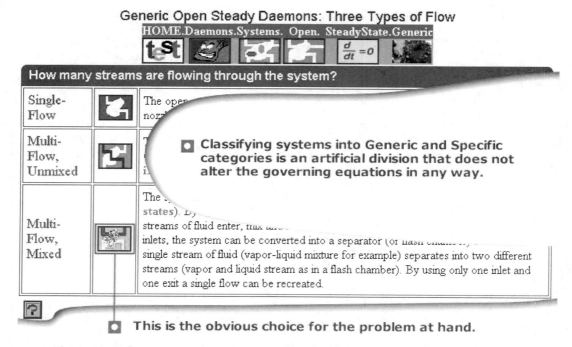

Generic Open Steady Daemons: Three Types of Flow

HOME.Daemons.Systems. Open. SteadyState.Generic

How many streams are flowing through the system?

Single-Flow	The open... nozz...
Multi-Flow, Unmixed	...
Multi-Flow, Mixed	The... states)... streams of fluid enter, mix an... inlets, the system can be converted into a separator (or flash chamb...) single stream of fluid (vapor-liquid mixture for example) separates into two different streams (vapor and liquid stream as in a flash chamber). By using only one inlet and one exit a single flow can be recreated.

🔲 Classifying systems into Generic and Specific categories is an artificial division that does not alter the governing equations in any way.

🔲 This is the obvious choice for the problem at hand.

■ *Balance Equations Ex. #1:* **The Systems.Open.Steady.Generic Page**

Nitrogen at 500 kPa and 50 deg-C enters a mixing... The balance equations remain unchanged whether a system is generic or application-specific.

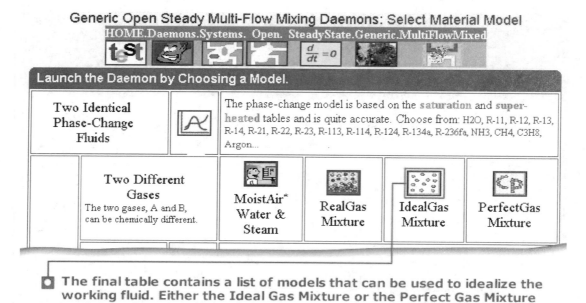

Generic Open Steady Multi-Flow Mixing Daemons: Select Material Model

HOME.Daemons.Systems. Open. SteadyState.Generic.MultiFlowMixed

Launch the Daemon by Choosing a Model.

| Two Identical Phase-Change Fluids | | The phase-change model is based on the saturation and superheated tables and is quite accurate. Choose from: H2O, R-11, R-12, R-13, R-14, R-21, R-22, R-23, R-113, R-114, R-124, R-134a, R-236fa, NH3, CH4, C3H8, Argon... |
| Two Different Gases The two gases, A and B, can be chemically different. | MoistAir* Water & Steam | RealGas Mixture | IdealGas Mixture | PerfectGas Mixture |

🔲 The final table contains a list of models that can be used to idealize the working fluid. Either the Ideal Gas Mixture or the Perfect Gas Mixture model can be used for the gaseous nitrogen and helium and their mixture, the working fluids of this problem.

■ *Balance Equations Ex. #1:* **The Systems.Open.Steady.Generic.Multi-Flow Mixed Page**

Nitrogen at 500 kPa and 50 deg-C enters a mixing... The final form of the balance equations can be found inside the Analysis panel of the daemons. To launch a daemon, a model for the working fluid must be chosen.

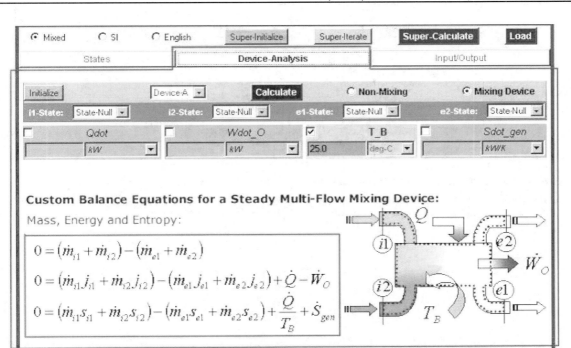

Balance Equations Ex. #1: The Balance Equations as Part of the Daemon

Nitrogen at 500 kPa and 50 deg-C enters a mixing... Although the equations allow for two exits, either of those can be plugged by leaving the state at "State-Null".

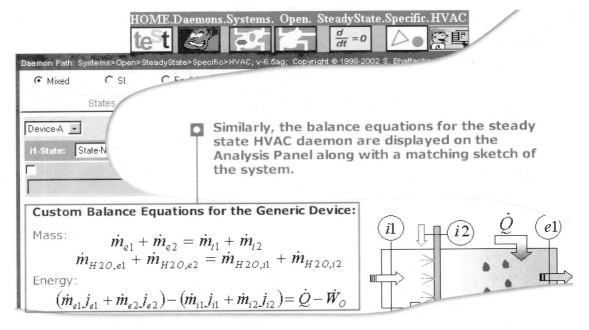

Similarly, the balance equations for the steady state HVAC daemon are displayed on the Analysis Panel along with a matching sketch of the system.

Balance Equations Ex. #1: Other Daemons

Nitrogen at 500 kPa and 50 deg-C enters a mixing... Here is another example with the HVAC daemon.

The step-by-step approach illustrated in this example can be applied to system problems of any kind to simplify the governing balance equations. Decoupling of the state relations from the balance equations is a sound practice in manual problem solving. Likewise in TEST, the balance equations are established before a material model is selected and the daemon is launched.

If you detect an error or simply have a suggestion, write to the author directly using the Comments page. Help make TEST a better software.

Balance Equations Ex. #1: TEST-Map vs. Systematic Approach

Nitrogen at 500 kPa and 50 deg-C enters a mixing... Although the TEST-Map provides a shortcut, there is no substitute for systematic simplification of the governing equation for a given problem.

Generic Closed Systems

All closed-systems problems encountered in the first half of most thermo textbooks are considered generic in TEST. The application-specific daemons are built upon these generic daemons.

Closed-system problems are subdivided into **Closed-Process** and **Closed-Steady** categories. The Closed-Process daemons build upon the Volume State daemons by adding the **Process-Analysis** and **Exergy** panels on top of the **States** panel. The beginning and end of a process are identified by the anchor states, the **b-** and the **f-states**, in the Process-Analysis panel of the **Uniform Closed-Process** daemons. The work transfer is divided into two quantities, W_B(W_B) for boundary work and W_O(W_O) for all other types of work, such as electric and shaft work. Once the anchor states are calculated, import them onto the process panel, and enter the known process variables. Note that the boundary work, W_B(W_B), is calculated by the daemon automatically whenever possible. The **Super-Calculate** button iterates between the balance and state equations to produce the desired answers. Alternatively, after the balance equations are solved on the Process-Analysis panel, you can switch to the States panel and complete the calculation of an unfinished state if new information (mdot, e or s) is posted there (marked by the gray background color) from the solution of the balance equations. Once a process is successfully analyzed, the exergy analysis is rather simple: Evaluate the dead state, import it to the **Exergy** panel, and **Calculate** all the exergy-related variables. Example 1 illustrates the Single-Flow daemon. The first two examples illustrate this daemon through screen shots.

For non-uniform systems, where more than one state is necessary to represent the composite nature of the system at a given time, the Non-Uniform daemons allow a composite begin state through **bA-** and **bB-states** and a composite finish state through **fA-** and **fB-states**. The Non-Uniform daemons come in two flavors, mixing vs. non-mixing types. They are illustrated in Examples 3 and 4. Finally, cycles are treated as closed steady systems for overall analysis, and the **Closed Steady** daemon, a stand-alone daemon customized for heat engines, heat pumps, and refrigerators, is illustrated in Example 5.

More detailed examples can be found in the Tutorial and in Chapters 3, 5, 6, and 7 of the Archive. The TEST-Codes in the Archive can be used to readily reproduce a solution in the classroom for the purpose of parametric studies.

Example: A piston–cylinder device contains 0.1 m^3 of air (ideal gas) at 300 kPa and 300 deg C. The system is now cooled at constant pressure until the temperature drops to 30 deg C. (a) Determine the heat transfer. (b) How would your answer change if the gas were CO_2 instead? (c) What if the perfect gas model is used? (d) If the ambient temperature is 25 deg C, determine the drop in exergy.

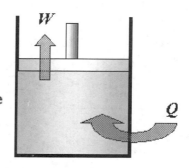

◻ **Solution Procedure: Solve the problem using the Generic, Closed-Process, Ideal Gas daemon. Save the TEST-Code generated by the Super-Calculate operation. Use this TEST-Code to regenerate the solution in the Perfect Gas daemon.**

■ *Closed Process Ex #1*: **Problem Description**

Closed Process Daemons: Generic and Specific Problems

HOME.Daemons.Systems.Closed.Process

te∘t   🐱   🧩   🐟 $_b\!\int^f$

Does the problem involve a special topic?		
Page	**Icon**	**Description**
Generic		Closed Process problems, except those involving special topics listed below, are handled here. Generally such problems are encountered in the first half of most thermodynamics textbooks.
Specific		Closed Processes involving special topics such as reciprocating cycles, combustion and psychrometry are handled in this branch.

?

Copyright 1998–2002: <u>Subrata Bhattacharjee</u>

◻ **The simplification tables, such as this one, lead an user through a series of assumptions necessary to launch the right daemon.**

■ *Closed Process Ex #1*: **Systematic Daemon Launch.**
A piston–cylinder device contains 0.1 m^3 of air... Click on the daemon link on the Task Bar to bring up the daemon page. Follow the directions in the simplification table.

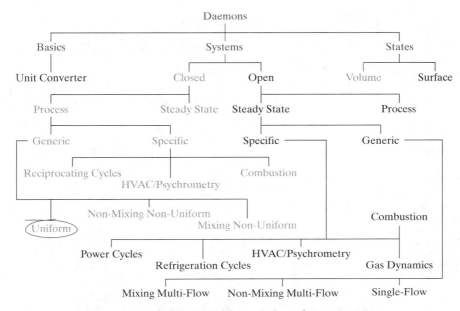

***Closed Process Ex. #1*: Simplify the Problem Using the TEST-Map**

A piston–cylinder device contains 0.1 m^3 of air... Use systematic navigation or the TEST-map to get to the right daemon page, Systems.Closed.Process.Generic.Uniform, for this problem.

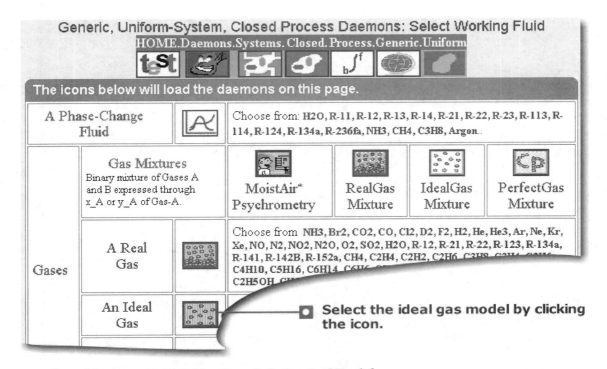

Select the ideal gas model by clicking the icon.

■ ***Closed Process Ex. #1*: Idealize—Select a Gas Model**

A piston–cylinder device contains 0.1 m^3 of air... Air, a mixture of O2 and N2 at a fixed proportion, can be treated as a pure gas. Let us select the ideal gas model.

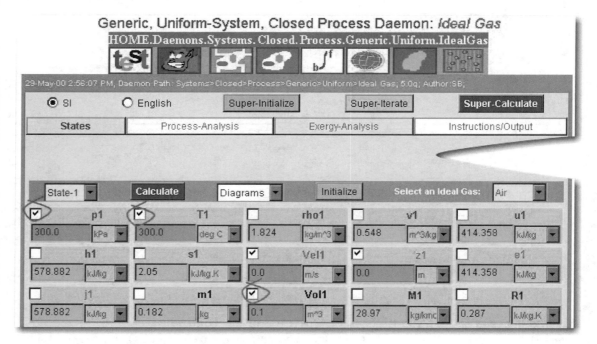

■ *Closed Process Ex. #1*: Evaluate the b(egin)-State

A piston–cylinder device contains 0.1 m^3 of air . . . Choose State-1. Air is the default fluid. Enter the known variables and Calculate.

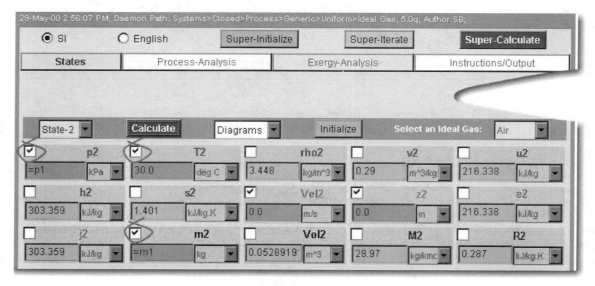

■ *Closed Process Ex. #1*: Evaluate the f(inish)-State

A piston–cylinder device contains 0.1 m^3 of air. . . Choose State-2 as the f-State. Enter the known variables (use algebraic expressions as much as possible) and Calculate.

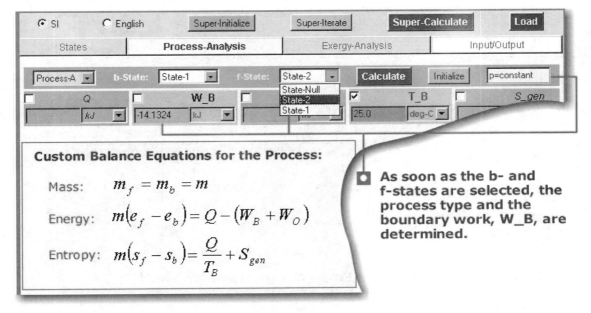

As soon as the b- and f-states are selected, the process type and the boundary work, W_B, are determined.

Custom Balance Equations for the Process:

Mass: $m_f = m_b = m$

Energy: $m(e_f - e_b) = Q - (W_B + W_O)$

Entropy: $m(s_f - s_b) = \dfrac{Q}{T_B} + S_{gen}$

■ *Closed Process Ex. #1*: **Analyze the Process**

A piston–cylinder device contains 0.1 m^3 of air. . . Switch to the analysis window by clicking the Process-Analysis tab. Load State-1 and State-2 as the b- and f-States. The boundary work calculated can be overwritten if necessary.

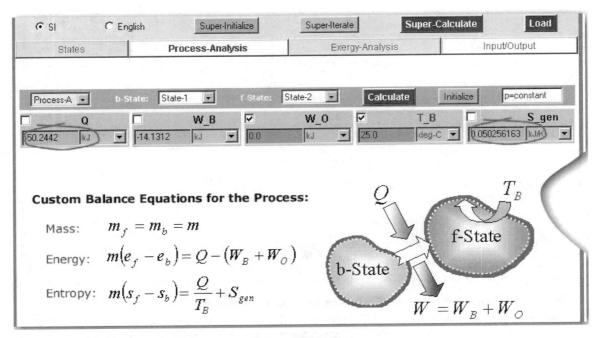

Custom Balance Equations for the Process:

Mass: $m_f = m_b = m$

Energy: $m(e_f - e_b) = Q - (W_B + W_O)$

Entropy: $m(s_f - s_b) = \dfrac{Q}{T_B} + S_{gen}$

$$W = W_B + W_O$$

■ *Closed Process Ex. #1*: **Heat Transfer Evaluated**

A piston–cylinder device contains 0.1 m^3 of air. . . Enter W_O = 0 and Calculate to determine Q. Note that the entropy generation is also calculated based on the default boundary temperature of 25 deg-C.

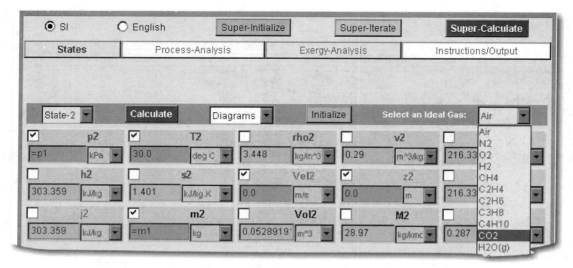

■ *Closed Process Ex. #1*: Parametric Study—Change the Working Fluid

How would your answer change if the gas is CO2 instead? Go back to the States panel and choose CO2 from the ideal gas selector.

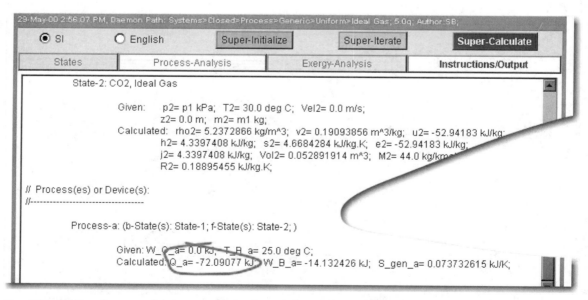

■ *Closed Process Ex. #1*: Super-Calculate for a Global Update

How would your answer change if the gas is CO2 instead? Click the Super-Calculate button. All the panels are updated. The new answer can also be found on this solution report.

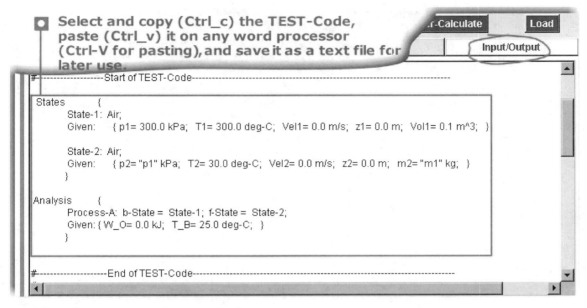

Select and copy (Ctrl_c) the TEST-Code, paste (Ctrl_v) it on any word processor (Ctrl-V for pasting), and save it as a text file for later use.

```
#----------------Start of TEST-Code-----------------------------------------

States     {
    State-1: Air;
    Given:    { p1= 300.0 kPa;  T1= 300.0 deg-C;  Vel1= 0.0 m/s;  z1= 0.0 m;  Vol1= 0.1 m^3;  }

    State-2: Air;
    Given:    { p2= "p1" kPa;  T2= 30.0 deg-C;  Vel2= 0.0 m/s;  z2= 0.0 m;  m2= "m1" kg;  }
    }

Analysis    {
    Process-A: b-State =  State-1;  f-State =  State-2;
    Given: { W_O= 0.0 kJ;  T_B= 25.0 deg-C;  }
    }

#----------------End of TEST-Code-----------------------------------------
```

Closed Process Ex. #1: TEST-Codes

A piston–cylinder device contains 0.1 m^3 of air. . . The TEST-Codes, generated by the Super-Calculate, can be used in several ways.

Closed Process Ex. #1: Using the TEST-Codes—Switch the Gas Model

What if the perfect gas model is used? To change the model, go back to the model selection page and launch the perfect gas daemon.

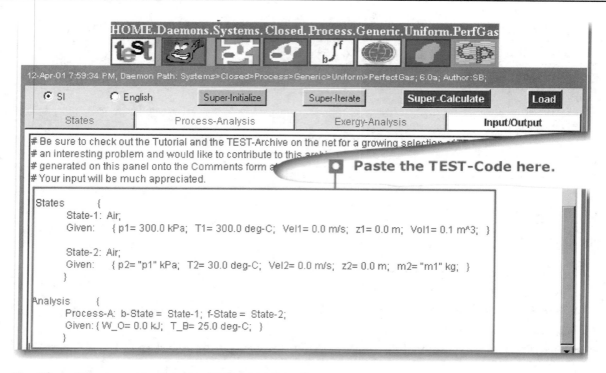

Closed Process Ex. #1: Paste the TEST-Codes

What if the perfect gas model is used? On the I/O panel of the perfect gas daemon, paste the TEST-Codes generated by the previous solution.

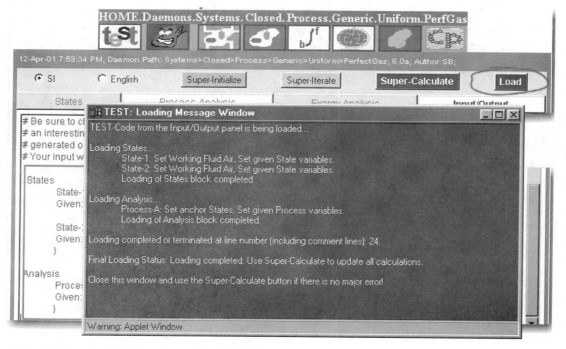

Closed Process Ex. #1: Load the TEST-Codes

What if the perfect gas model is used? The Load operation compiles the code. Close the Message Window, which displays the loading status and error messages, if any.

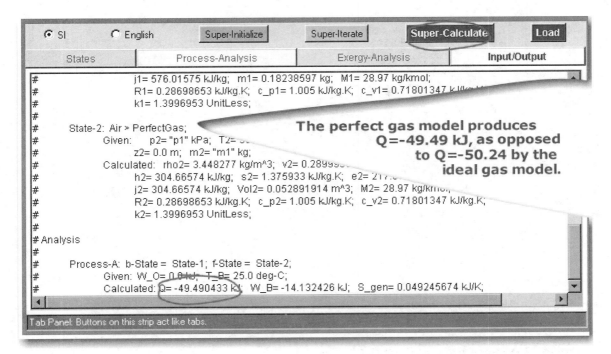

Inside the Input/Output panel:

```
#                j1= 576.01575 kJ/kg;  m1= 0.18238597 kg;  M1= 28.97 kg/kmol;
#                R1= 0.28698653 kJ/kg.K;  c_p1= 1.005 kJ/kg.K;  c_v1= 0.71801347 kJ/kg.K;
#                k1= 1.3996953 UnitLess;
#
#        State-2: Air > PerfectGas;
#              Given:    p2= "p1" kPa;  T2= ...
#                        z2= 0.0 m;  m2= "m1" kg;
#              Calculated:  rho2= 3.448277 kg/m^3;  v2= 0.2899...
#                        h2= 304.66574 kJ/kg;  s2= 1.375933 kJ/kg.K;  e2= 21...
#                        j2= 304.66574 kJ/kg;  Vol2= 0.052891914 m^3;  M2= 28.97 kg/kmol;
#                        R2= 0.28698653 kJ/kg.K;  c_p2= 1.005 kJ/kg.K;  c_v2= 0.71801347 kJ/kg.K;
#                        k2= 1.3996953 UnitLess;
#
# Analysis
#
#        Process-A:  b-State = State-1;  f-State = State-2;
#              Given:  W_O= 0.0 kJ;  T_B= 25.0 deg-C;
#              Calculated: Q= -49.490433 kJ;  W_B= -14.132426 kJ;  S_gen= 0.049245674 kJ/K;
```

The perfect gas model produces Q=-49.49 kJ, as opposed to Q=-50.24 by the ideal gas model.

Closed Process Ex. #1: Super-Calculate to Instantly Generate the Complete Solution

What if the perfect gas model is used? Super-Calculate updates all calculations and produces a Q, which is slightly smaller than the Q produced by the ideal gas model. (Why?)

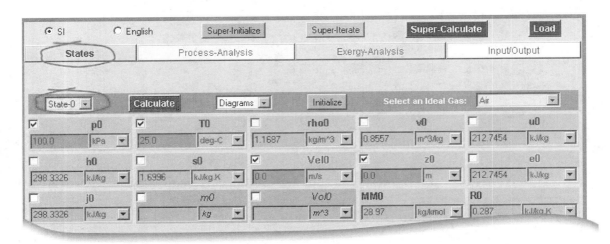

For exergy calculations, designate state-0 as the dead state, and Calculate it based on the ambient pressure and temperature.

Closed Process Ex. #1: Evaluate the Dead State

Determine the change in exergy. Select State-0 as the dead state, enter p0 and T0, and Calculate.

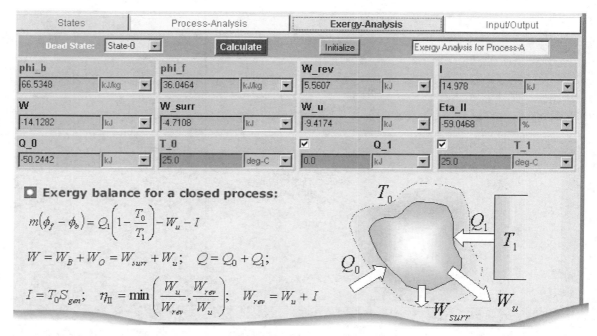

Closed Process Ex. #1: The Exergy Panel

Determine the loss of exergy. Switch to the exergy panel, load State-0, and Calculate all the exergy-related variables. The daemon allows a second thermal reservoir ($Q_1 = 0$ in this problem). The loss of exergy, from the balance equation, is $I + W_u = 14.98 - 9.42 = 5.56$ kJ.

```
###################################################################
# To regenerate this solution, copy the following TEST-Code onto the I/O panel of the
# ...Closed.Process.Generic.Uniform.IdealGas daemon.
# and click the Load and Super-Calculate buttons.
#--------------------Start of TEST-Code------------------------------------------------
----

 States {
         State-1:  Air;
         Given:     { p1 = 300.0 kPa;  T1 = 300.0 deg-C;  Vel1 = 0.0 m/s;
                       z1 = 0.0 m;  Vol1 = 0.1 m^3;  }

         State-2:  Air;
         Given:     { p2 = "p1" kPa;  T2 = 30.0 deg-C;  Vel2 = 0.0 m/s;  z2 = 0.0 m;  }
 }

 Analysis {
         Process-A:  b-State = State-1;  f-State = State-2;
         Given: { W_O = 0.0 kJ;  T_B = 25.0 deg-C;  }
 }
#--------------------End of TEST-Code---------------------------------------------------
-----
```

Closed Process Ex. #1: TEST-Code to Regenerate this Solution

Example: A rigid container of volume 1.0 m3 contains ammonia at 100 kPa and 90% quality. If 200 kJ of heat is transferred to the tank, determine the final pressure and temperature.

What-If Scenario: (a) How would the answers change if the amount of heat transfer were 400 kJ?

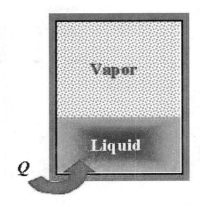

□ **Solution Procedure:** Simplify the problem as a closed, uniform system undergoing a process. Launch the closed process daemon, evaluate the begin and finish states, load the states into the analysis panel, enter the known process variables, and Super-Calculate the answers.

■ *Closed Process Ex. #2:* **Problem Description**

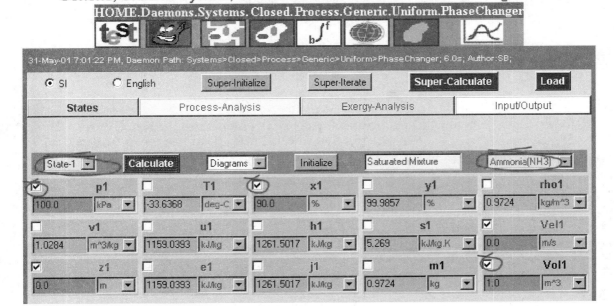

■ *Closed Process Ex. #2:* **Evaluate the b(egin)-State**

A rigid container of volume 1 m^3 contains ammonia... The simplification process for this problem is the same as in the last example. Pick the phase-change model to represent ammonia. Evaluate the begin-state, State-1, from the given conditions (p, x, and Vol).

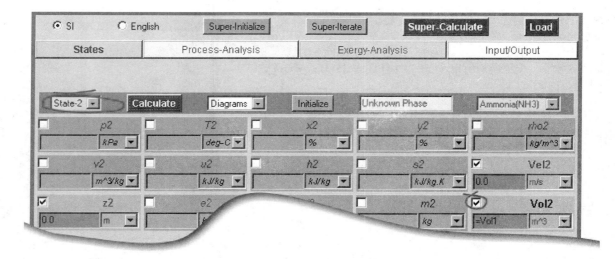

▣ **Very little is known about State-2, except for the volume. The balance equations will produce the missing links, m2 (obvious) and e2.**

■ *Closed Process Ex. #2:* **Evaluate the Finish-State**

A rigid container of volume 1 m^3 contains ammonia. . . Select State-2 as the f-state. Enter Vol, the only known variable, and partially evaluate the state.

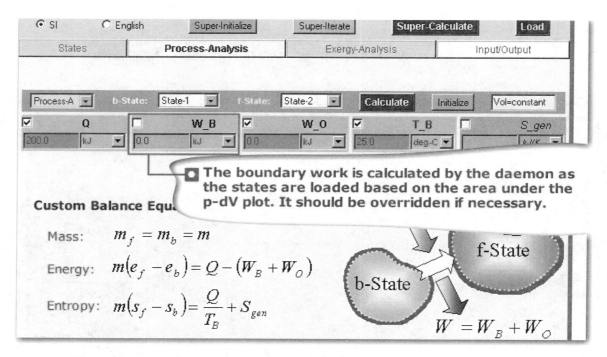

▣ **The boundary work is calculated by the daemon as the states are loaded based on the area under the p-dV plot. It should be overridden if necessary.**

Custom Balance Equ...

Mass: $m_f = m_b = m$

Energy: $m(e_f - e_b) = Q - (W_B + W_O)$

Entropy: $m(s_f - s_b) = \dfrac{Q}{T_B} + S_{gen}$

f-State

b-State

$W = W_B + W_O$

■ *Closed Process Ex. #2:* **Solve the Balance Equations**

A rigid container of volume 1 m^3 contains ammonia. . . Load the b- and f-states, enter Q and W_O, and Calculate. The state variables calculated from the balance equations (m_f and e_f) are posted back to the f-state, State-2. Switch to the States panel.

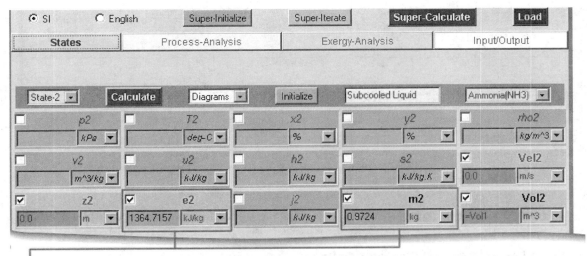

■ Solutions from the mass and energy balance equations, m2 and e2, are
posted back to State-2. Calculate or, even better, Super-Calculate
finishes up the job.

■ *Closed Process Ex. #2*: **Analysis Results Posted**

A rigid container of volume 1 m^3 contains ammonia... With m2 and e2 posted by the Analysis panel, State-2
is now ready to be evaluated. Calculate can evaluate the state, but Super-Calculate is preferable because it
updates all panels at once.

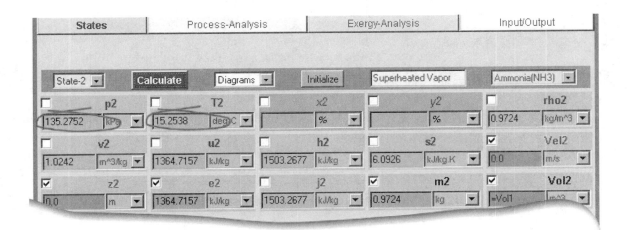

■ Note that while a manual solution for this state (from v2 and u2)
involves iterations, the TEST procedure remains identical, irrespective of
the difficulty of finding a state.

■ *Closed Process Ex. #2*: **Finish-State Evaluated**

A rigid container of volume 1 m^3 contains ammonia... State-2 is now completely evaluated.

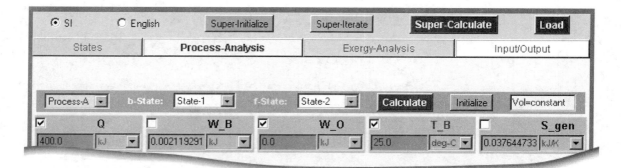

> **Note that the S_gen value is based on a default boundary temperature, which may not be correct in a situation where a system is heated.**

Closed Process Ex. #2: Parametric Study—Effect of Heat Transfer

A rigid container of volume 1 m^3 contains ammonia... For the parametric study, change Q to the new value, Calculate (or press Enter) and Super-Calculate.

```
############################################################################
# To regenerate this solution, copy the following TEST-Code onto the I/O panel of the
# ..Closed.Process.Generic.Uniform.PhaseChanger.
# and click the Load and Super-Calculate buttons.
#--------------------Start of TEST-Code----------------------------------------
----

States {
State-1: Ammonia(NH3);
Given:      { p1 = 100.0 kPa;  x1 = 90.0 %;  Vel1 = 0.0 m/s;  z1 = 0.0 m;  Vol1 = 1.0 m^3;  }

State-2: Ammonia(NH3);
Given:      { Vel2 = 0.0 m/s;  z2 = 0.0 m;  Vol2 = "Vel1" m^3;  }
}

Analysis {
Process-A: b-State =  State-1;  f-State =  State-2;
Given: { Q = 400.0 kJ;  W_O = 0.0 kJ;  T_B = 25.0 deg-C;  }
}
```

Closed Process Ex. #2: TEST Code to Regenerate the Visual Solution

A 2500-gallon insulated rigid tank containing H2 at 68 deg F 29 psia is connected to another 50-gallon insulated rigid tank containing CO_2 at 200 deg F and 90 psia. The valve is opened and the system is allowed to reach thermal equilibrium. Determine the final pressure, heat transfer and the entropy generation. Assume constant specific heats.

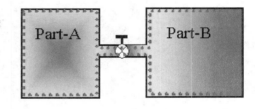

What-If Scenario: Determine the heat transfer necessary to bring the mixture back to the ambient temperature of 65 deg F.

> ☐ **Solution Procedure:** Simplify the problem as a closed, non-uniform system undergoing a mixing process. Launch the closed process mixing daemon, evaluate the begin and finish states, load the states into the analysis panel, enter the known process variables, and Super-Calculate the answers.

■ *Closed Process Ex. #3*: **Problem Description**

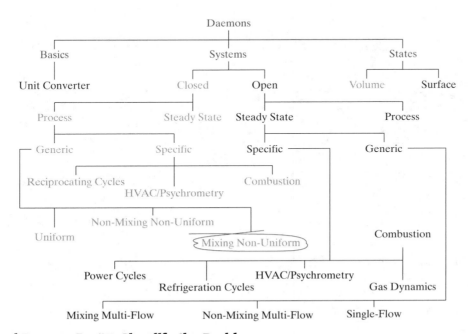

■ *Closed Process Ex. #3*: **Simplify the Problem**

A 2500-gallon insulated rigid tank containing. . . Here we use the TEST-Map to simplify and reach the appropriate daemon page, . . . Closed.Process.Generic.Non-Uniform-Mixing.

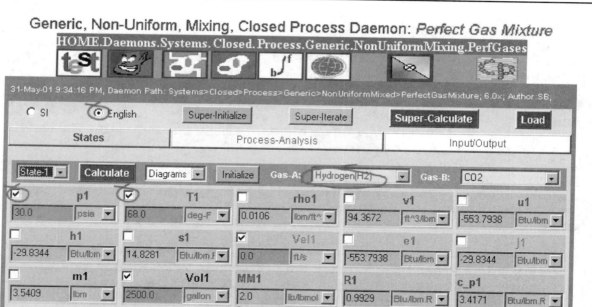

Closed Process Ex. #3: Evaluate the Composite b(egin)-States

A 2500-gallon insulated rigid tank containing. . . Two states are necessary to describe the non-uniform begin state. Notice how x_A = 1 or 0 converts the mixture into one of its pure components.

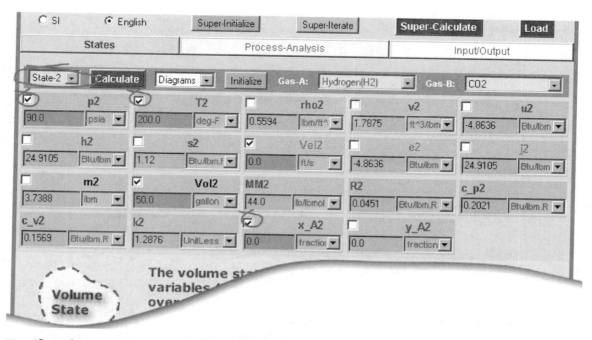

Closed Process Ex. #3: Evaluate the b(egin) States

A 2500-gallon insulated rigid tank containing. . . Here x_A = 0 converts the mixture into pure CO2. State-1 and State-2 form the composite begin state.

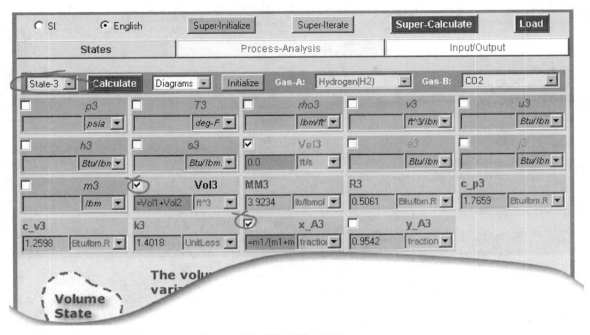

Closed Process Ex. #3: Evaluate the f(inish)-State

A 2500-gallon insulated rigid tank containing... The f-state has a mixture. Note how the mass fraction, x_A, is expressed in terms of m1 and m2. We did not enter m3 = m1 + m2, leaving it for the mass balance equation to deduce and post on this panel.

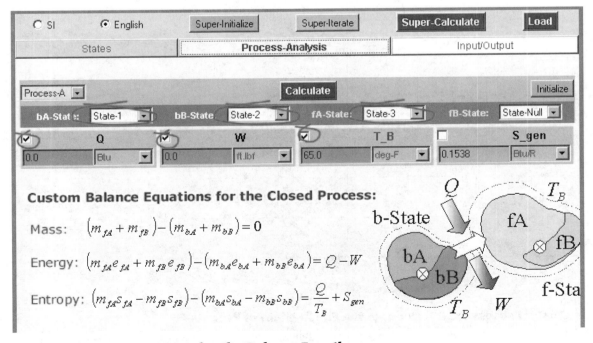

Closed Process Ex. #3: Solve the Balance Equations

A 2500-gallon insulated rigid tank containing... On the Analysis panel, load the bA- and bB-states and the single f-state (as fA or fB). Enter Q, W, and T_B, and Calculate. Super-Calculate updates all panels.

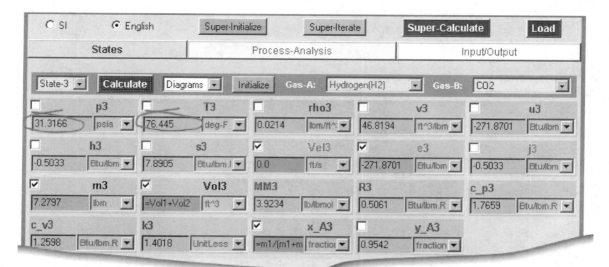

■ The process analysis produces m3 and e3, which are posted back to State-3 before Super-Calculate evaluates the complete state.

■ *Closed Process Ex. #3:* **State-3 Is Found**

A 2500-gallon insulated rigid tank containing. . .The final pressure and temperature are determined as part of State-3, which is completely evaluated now that m3 and e3 have been posted by the Analysis panel.

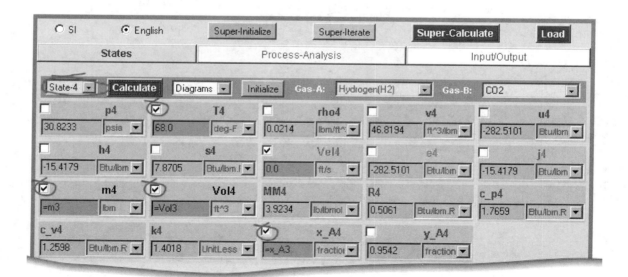

■ *Closed Process Ex. #3:* **Evaluate the Finish-State of Process-B**

A 2500-gallon insulated rigid tank containing. . .The cooling down of the mixture is another closed process, except the working fluid is now uniform during the entire process. Evaluate the finish-state, State-4.

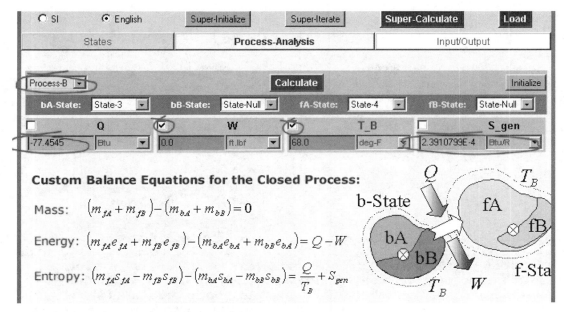

Closed Process Ex. #3: Analyze Process B

Closed Process Ex. #3: A 2500-gallon insulated rigid tank containing. . . Select Process-B, load the single b-State and f-State, and enter W and T_B. Calculate produces the desired heat transfer.

```
######################################################################
# To regenerate this solution, copy the following TEST-Code onto the I/O panel of the
# ..Closed.Process.Generic.MixingNon-Uniform.PerfectGases
# and click the Load and Super-Calculate buttons.
#-------------------Start of TEST-Code-----------------------------------
----

States {
State-1:  Hydrogen(H2), CO2;
Given:      { p1= 30.0 psia;  T1= 68.0 deg-F;  Vel1= 0.0 ft/s;  Vol1= 2500.0 gallon;  x_A1=
1.0 fraction;  }

State-2: Hydrogen(H2), CO2;
Given:      { p2= 90.0 psia;  T2= 200.0 deg-F;  Vel2= 0.0 ft/s  Vol2= 50.0 gallon;  x_A2 = 0.0
fraction;  }

State-3: Hydrogen(H2), CO2;
Given:      { Vel3= 0.0 ft/s;  Vol3 = "Vol1+Vol2" ft^3;  x_A3 = "m1/(m1+m2)" fraction;  }

State-4: Hydrogen(H2), CO2;
Given:      { T4= 20.0 deg-C;  Vel4= 0.0 m/s;  Vol4="Vol3" m^3;  x_A4 = "x_A3"
fraction;  }
}

Analysis {
Process-A: b-State =  State-1, State-2;  f-State =  State-3;
Given: { Q = 0.0 Btu;  W= 0.0 ft.lbf;  T_B = 65.0 deg-F;  }

Process-B: b-State = State-3;  f-State = State-4;
Given: { W= 0.0 kJ;  T_B= 25.0 deg-C;  }
}
```

Closed Process Ex. #3: TEST-Code to Regenerate the Visual Solution

Example: An unknown mass of iron at 80 deg-C is dropped into an insulated tank that contains 0.1 m3 of liquid water at 20 deg-C. Meanwhile, a paddle wheel driven by a 200 W motor is used to stir the water. When equilibrium is reached after 20 min, the final temperature is 25 deg-C. Determine the mass of the iron block. How would the answer change if the final temperature is 30 deg-F?

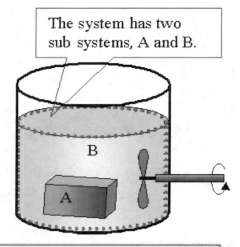

The system has two sub systems, A and B.

⬛ **Solution Procedure:** Simplify the problem as a closed, non-uniform system undergoing a process. Launch the appropriate closed process daemon, evaluate the begin and finish states, load the states in the analysis panel, enter the known process variables and Super-Calculate the answers.

◼ *Closed Process Ex. #4*: **Problem Description**

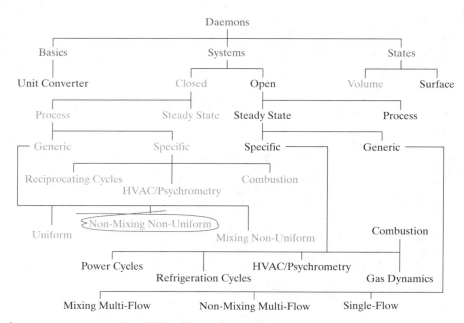

◼ *Closed Process Ex. #4*: **Simplify Using the TEST-Map**

An unknown mass of iron at 80 deg-C. . . The appropriate simplification for this non-mixing process is. . . Closed.Process.Generic.Non-Uniform-Non-Mixing.

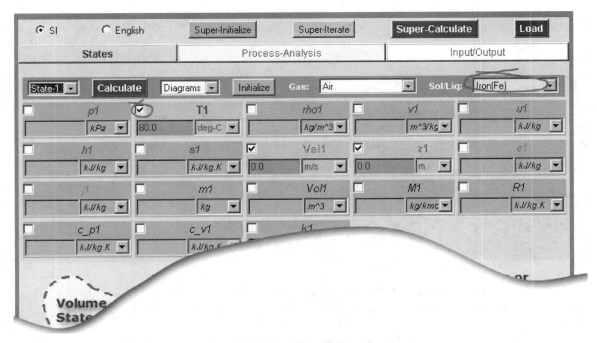

■ *Closed Process Ex. #4* : Evaluate the Composite Begin-States

An unknown mass of iron. . . The begin-state is composed of two states. Select Iron as the working substance for State-1. Only T1 is known at this point.

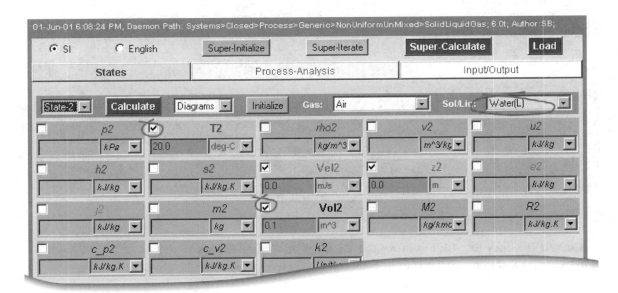

■ *Closed Process Ex. #4*: Evaluate the Composite Begin-States

An unknown mass of iron. . . Select Water as the working substance for State-2 and enter the known properties.

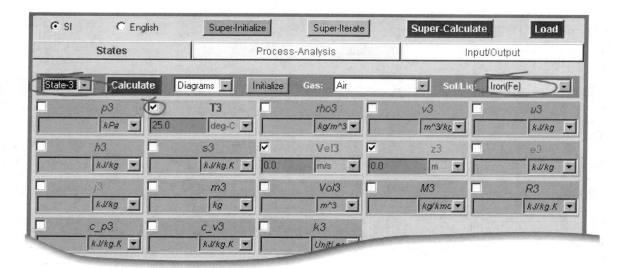

Closed Process Ex. #4: Evaluate the Composite Finish-States

An unknown mass of iron...The finish-state is also non-uniform. The mass of iron is unknown, but the temperature is known.

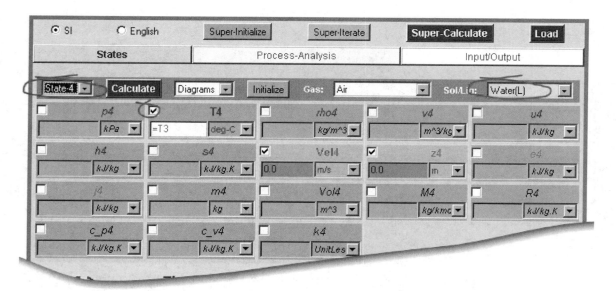

Closed Process Ex. #4: Evaluate the Composite Finish-States

An unknown mass of iron...Select Water(L) and enter the known properties for the fB-State.

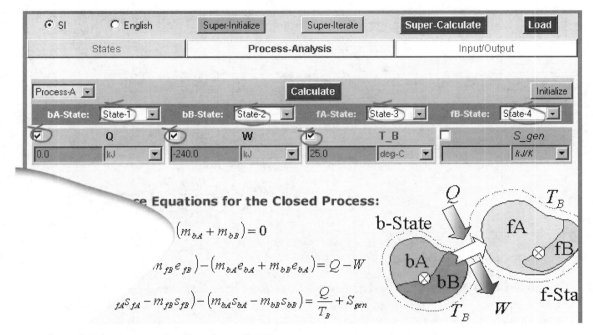

■ *Closed Process Ex. #4*: **Analyze the Process**

An unknown mass of iron. . . Load the four b- and f-states, enter the known Q and W, and Calculate. Super-Calculate produces the desired answers in the appropriate panels.

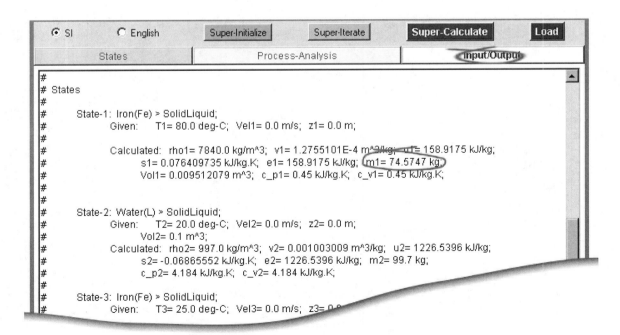

■ *Closed Process Ex. #4*: **Solution Report**

An unknown mass of iron. . . The answers can be found in the appropriate panels or on the I/O panel, where a detailed solution report is generated by Super-Calculate.

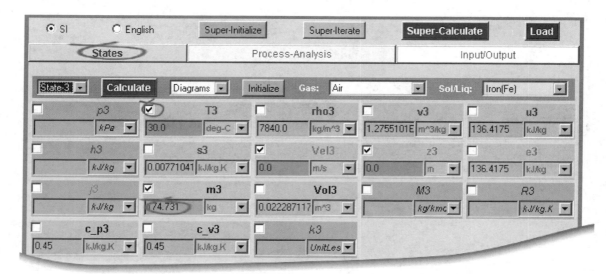

Closed Process Ex. #4: Parametric Study—Vary the Final Temperature

An unknown mass of iron. . . Change T3 to 30 deg-C, press the Enter key (or Calculate), and Super-Calculate. The new mass corresponding to this new input is found.

```
###########################################################################
# To regenerate this solution, copy the following TEST-Code onto the I/O panel of the
# ..Closed.Process.Generic.NonUniformUnmixed.Solid-Liquid daemon
# and click the Load and Super-Calculate buttons.
#-------------------Start of TEST-Code----------------------------------------------------
----

States {
State-1:  Iron(Fe);
Given:      { T1= 80.0 deg-C;  Vel1= 0.0 m/s;  z1= 0.0 m;  }

State-2: Water(L);
Given:      { T2= 20.0 deg-C;  Vel2= 0.0 m/s;  z2= 0.0 m; Vol2= 0.1 m^3;  }

State-3:  Iron(Fe);
Given:      { T3T= 30.0 deg-C;  Vel3= 0.0 m/s;  z3= 0.0 m;  }

State-4:  Water(L);
Given:      { T4= "T3" deg-C;  Vel4= 0.0 m/s;  z4= 0.0 m;  }

Analysis {
Process-A: b-State = State-1, State-2;  f-State = State-3, State-4;
Given: { Q= 0.0 kJ;  W= −240.0 kJ; T_B= 25.0 deg-C;  }
}
```

Closed Process Ex. #4: TEST-Code to Regenerate the Visual Solution

Example: On a cold night, a house is losing heat at a rate of 80,000 Btu/h. A reversible heat pump maintains the house at 70 deg F, while the outside temperature is 30 deg F.

Determine the heating cost for the night (8 hours), assuming the price of 10 cents/kWh for electricity. Also, determine the heating cost if resistance heating is used instead.

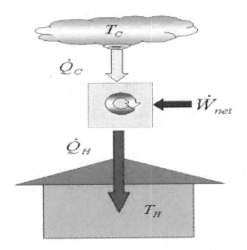

◘ **Solution Procedure:** Simplify the problem (Closed & Steady) and launch the appropriate daemon. Note that the problem involves only overall cycle quantities and, therefore, can be treated as a cycle problem. If details of the refrigeration cycles were involved, we would have chosen a different daemon (Open...Specific.Refrigeration).

■ *Closed System Ex. #5*: **Problem Description and Solution Algorithm**

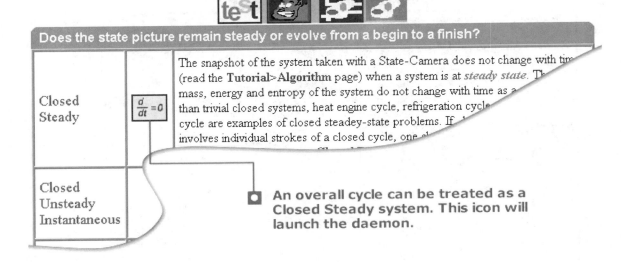

Closed System Daemons: Closed Process or Closed Steady Problems

HOME.Daemons.Systems.Closed

Does the state picture remain steady or evolve from a begin to a finish?

Closed Steady

$\frac{d}{dt} = 0$

The snapshot of the system taken with a State-Camera does not change with time (read the **Tutorial>Algorithm** page) when a system is at *steady state*. The mass, energy and entropy of the system do not change with time as a than trivial closed systems, heat engine cycle, refrigeration cycle cycle are examples of closed steadey-state problems. If involves individual strokes of a closed cycle, one

Closed Unsteady Instantaneous

◘ An overall cycle can be treated as a Closed Steady system. This icon will launch the daemon.

■ *Closed System Ex. #5*: **Systematic Daemon Launch**

On a cold night, a house ... Click on the daemon link on the Task Bar to bring up the daemon page. Follow the directions in the simplification table.

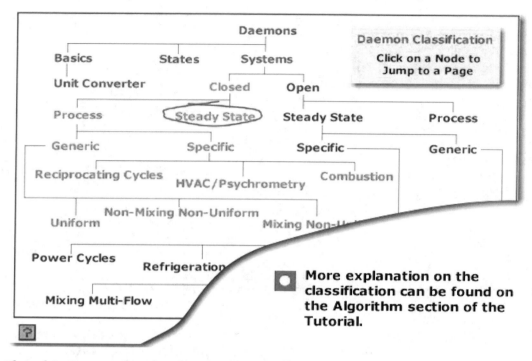

More explanation on the classification can be found on the Algorithm section of the Tutorial.

■ *Closed System Ex. #5*: Use the TEST-Map to Simplify

On a cold night, a house ... Use systematic navigation or the TEST-map to get to the right daemon page, Systems. Closed.Steady, for this problem.

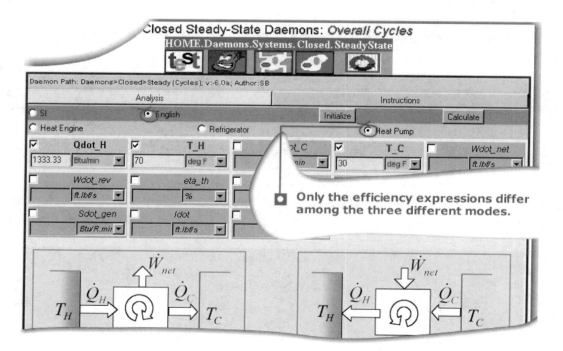

Only the efficiency expressions differ among the three different modes.

■ *Closed System Ex. #5*: Enter Known Values

On a cold night a house... Choose English units and Heat Pump radio buttons. Enter the temperature of the hot (inside) and cold (outside) reservoirs.

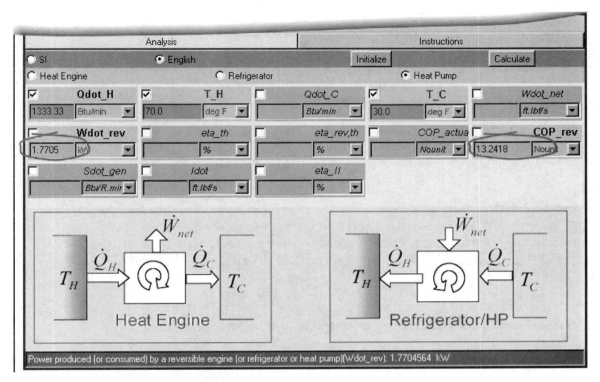

Closed System Ex. #5: Calculate

On a cold night, a house... Calculate produces the reversible power as 1.77 kW.

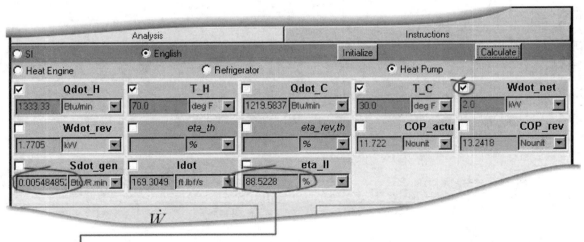

The second law efficiency measures the perfection of the heat pump by comparing its COP with the COP of a reversible (perfect) system pumping heat between the same reservoirs.

Closed System Ex. #5: The 2nd Law Efficiency

On a cold night a house... Enter the actual power as 2 kW and Calculate the second-law efficiency and the rate of entropy generation.

EXAMPLE: A 10 m2 brick wall separates two chambers at 500 K and 300 K respectively. If the rate of heat transfer is 0.5 kW/m2, determine the entropy generation rate in the wall due to the heat transfer. Assume the wall surface temperatures to be the same as the adjacent chamber temperatures. Also assume steady state.

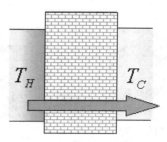

⬛ **Solution Procedure: Launch the Closed Steady daemon. The wall can be treated as a heat engine with no work output. Enter Qdot_H=Qdot_C=0.5*10=5 kW, T_H and T_C. Calculate the rate of entropy generation.**

■ *Closed System Ex. #6*: **Problem Description and Solution Algorithm**

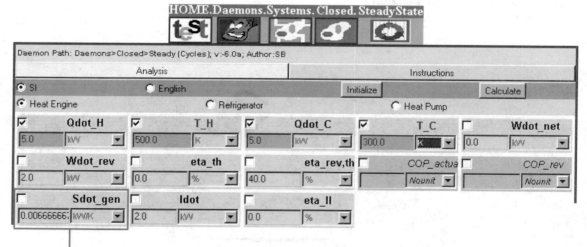

⬛ Note that the thickness of the wall has no effect on the entropy generation rate. Also note that the reversible work is 2 kW, which is also the rate of the exergy destruction (Idot), thereby, ensuring that the wall does not produce any net work.

■ *Closed System Ex. #6*: **Enter Known Values and Calculate**

A 10 m2 brick wall. . . Enter Qdot_H, Qdot_C, T_H, T_C and Wdot_net = 0. Calculate.

For more examples on this topic, check out the Tutorial > Daemons > Closed-Process and Closed-Steady pages. A large number of solutions and TEST-Codes can be found in Chapter-3, 5, 6 and 7 of the Archive, organized into fifteen chapters consistent with most textbooks.

If you detect an error, would like more solutions on this topic, or simply have a suggestion, write to the author directly using the Comments page. Help make TEST a better software.

Generic Open Systems

All open system problems encountered in the first half of most thermo text-books are considered generic in TEST. The application-specific daemons are built upon these generic daemons.

Open system problems are subdivided into **Open-Steady** and **Open-Process** categories. The Open Steady daemons build upon the Surface State daemons by adding the **Device-Analysis** and **Exergy** panels on top of the **States** panel. The inlet and exit states of a device are identified by the anchor states, the **i-** and the **e-state**, in the Device-Analysis panel of the **Single-Flow Open Steady** daemons. TEST introduces a new symbol, j, to represent the flow energy

$$j \equiv h + \frac{V^2}{2} + gz$$

absorbing the flow work into the convection terms. At steady state, the boundary work must be zero, leaving $\dot{W}_O$ (Wdot_O)(from shaft or electrical power) as the only means of work transfer. Once the anchor states are calculated, import them onto the device panel and enter the known device variables. The **Super-Calculate** button iterates between the balance and state equations to produce the desired answers. Alternatively, after the balance equations are solved on the Device-Analysis panel, you can switch to the States panel and complete the calculation of an unfinished state if new information (on mdot, j or s) is posted there (marked by the gray background color) from the solution of the balance equation. Once a device is successfully analyzed, the exergy analysis is rather simple. Evaluate the dead state, import it to the **Exergy** panel, and **Calculate** all the exergy-related variables. Examples 1 and 2 illustrate the Single-Flow daemon.

For multi-flow systems, where more than one state is necessary to represent the states of the multiple inlet and exit ports, the Multi-Flow daemon offers up to two inlet ports, identified by **i1-** and **i2-state**, and two exit ports, **e1-** and **e2-state**. With the use of a radio button, the internal connections among the ports can be manipulated, switching the device between mixing and non-mixing types. If a state is not imported to a port, it is considered plugged, thereby allowing a multi-flow device to be turned into a single-flow device, a mixing device, or a separator. These are illustrated in Examples 3 and 4.

The Open Process daemons, illustrated in Example 5, handle problems on charging and discharging. The Analysis panel combines the **b-** and **f-state** for the process with the **i-** and **e-state** of the possible inlet or exit port.

Example: Saturated liquid water steadily enters a valve at 400 psia, 10 ft/s, and 1 lbm/s and exits at a pressure of 70 psia. Neglecting any heat transfer, determine (a) what property remains constant between the inlet and exit; (b) the exit quality if the exit area is the same as the inlet area; (c) the exit area if the exit velocity is the same as the inlet velocity; (d) the rate of irreversibility in Btu/min if the ambient conditions are 77 deg F and 1 atm.

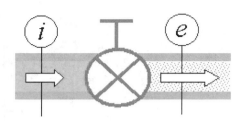

(e) How would the answers in parts (a) and (b) change if the velocity were neglected in this throttling process? (f) Convert the entire solution into SI units.

> ◻ **Solution Procedure: Simplify the problem as an open, generic, steady, single-flow problem with phase-change model for H$_2$O. Evaluate the i- and e-states, load them on the Analysis panel, enter Wdot_O and Qdot as zero, and Super-Calculate.**

■ *Open System Ex. #1*: **Problem Description**

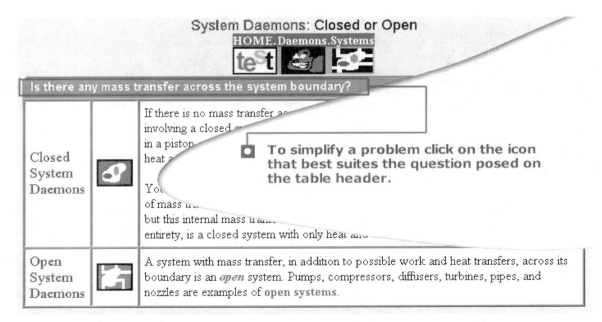

■ *Open System Ex. #1*: **Systematic Daemon Launch.**

Saturated liquid water steadily enters a valve.... Click on the daemon link on the Task Bar to bring up the daemon page. Follow the directions in the simplification table.

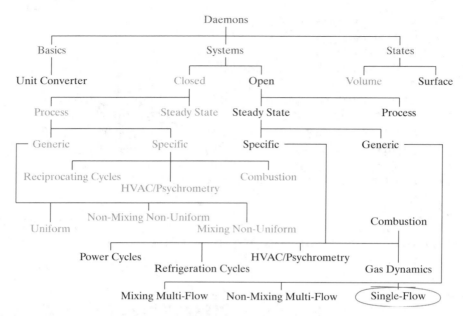

■ *Open System Ex. #1*: Simplify Using the TEST-Map

Saturated liquid water steadily enters a valve...The appropriate simplification for this problem takes us to the . . . Open.Steady.Generic.Single Flow page.

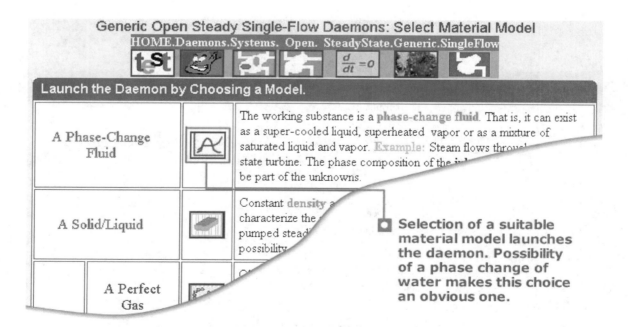

Selection of a suitable material model launches the daemon. Possibility of a phase change of water makes this choice an obvious one.

■ *Open System Ex. #1*: Idealization - Pick a Fluid Model

Saturated liquid water steadily enters a valve...The phase-change model is the obvious choice to entertain the possibility of phase change during the process.

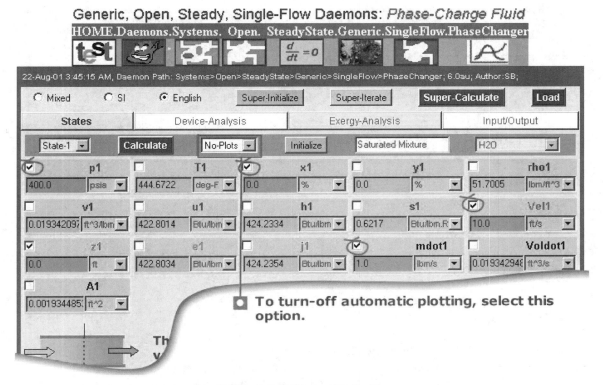

To turn-off automatic plotting, select this option.

■ *Open System Ex. #1*: Evaluate the Inlet State (State-1)

Saturated liquid water steadily enters a valve… H_2O is the default fluid. Select the English system and State-1, and enter p1, x1, Vel1, and mdot1.

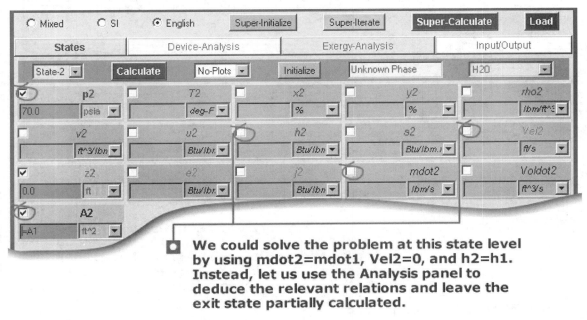

We could solve the problem at this state level by using mdot2=mdot1, Vel2=0, and h2=h1. Instead, let us use the Analysis panel to deduce the relevant relations and leave the exit state partially calculated.

■ *Open System Ex. #1*: Partially Evaluate the Exit State (State-2)

Saturated liquid water steadily enters a valve… Select State-2, enter p2 and A2, and Calculate. More information is required before the state can be fully evaluated.

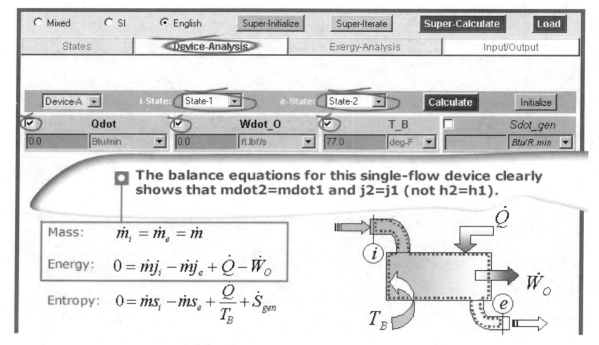

The balance equations for this single-flow device clearly shows that mdot2=mdot1 and j2=j1 (not h2=h1).

Mass: $\dot{m}_i = \dot{m}_e = \dot{m}$

Energy: $0 = \dot{m}j_i - \dot{m}j_e + \dot{Q} - \dot{W}_O$

Entropy: $0 = \dot{m}s_i - \dot{m}s_e + \dfrac{Q}{T_B} + \dot{S}_{gen}$

■ Open System Ex. #1: Analyze the Device

Saturated liquid water enters steadily a valve... On the Analysis panel, enter Qdot and Wdot_and Calculate. Note that friction in a valve is essential, so Sdot_gen must be significant.

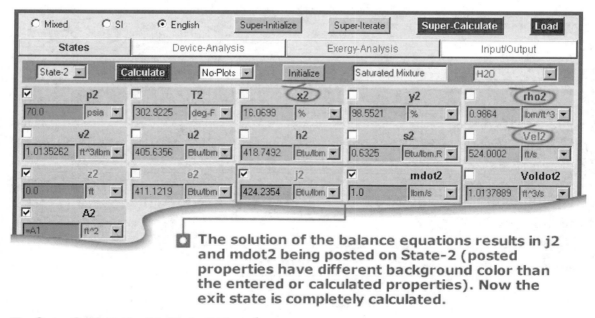

The solution of the balance equations results in j2 and mdot2 being posted on State-2 (posted properties have different background color than the entered or calculated properties). Now the exit state is completely calculated.

■ Open System Ex. #1: State-2 Found

Saturated liquid water enters steadily a valve... On State-2, the gray background of j2 and $$$ indicates that these variables have been imported. A Calculate or a Super-Calculate complete evaluates this state.

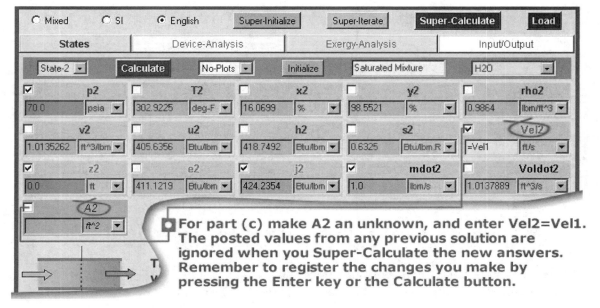

For part (c) make A2 an unknown, and enter Vel2=Vel1. The posted values from any previous solution are ignored when you Super-Calculate the new answers. Remember to register the changes you make by pressing the Enter key or the Calculate button.

■ **Open System Ex. #1: Modify the Exit Conditions**

Saturated liquid water enters steadily a valve. . . Change Vel2 and make A2 an unknown. Press enter to register the changes and Super-Calculate.

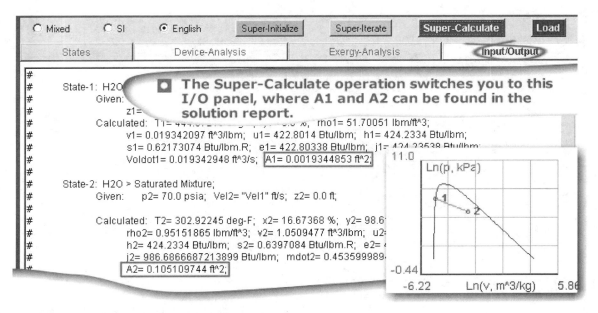

The Super-Calculate operation switches you to this I/O panel, where A1 and A2 can be found in the solution report.

■ **Open System Ex. #1: The Solution Report**

Saturated liquid water steadily enters a valve. . . You can also go back to the States panel to see the updated state visually.

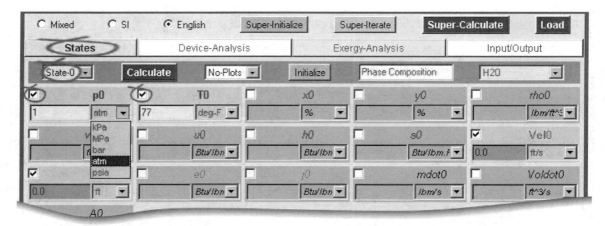

□ Go back to the States panel, start a new state, State-0, for the dead state, enter p0 and T0, and Calculate.

■ **Open System Ex. #1: The Dead State (State-0)**

Saturated liquid water steadily enters a valve. . . Note that the working fluid for the dead state is H2O, not air.

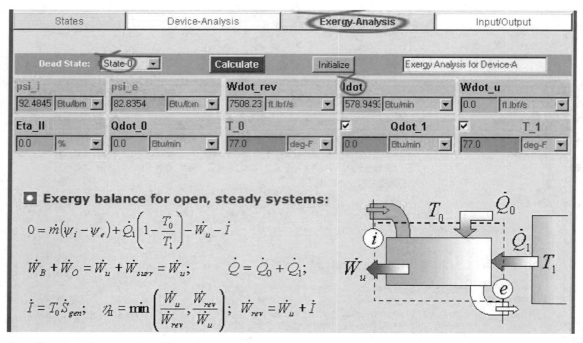

■ **Open System Ex. #1: The Exergy Panel**

Saturated liquid water steadily enters a valve. . . On the exergy panel, load the dead state and Calculate all the exergy-related variables. The second–law efficiency is zero because all the reversible work that could be produced is lost to irreversibilities.

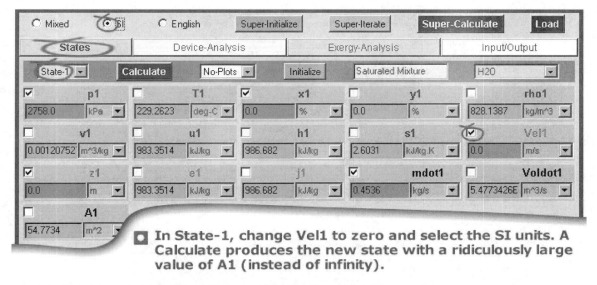

In State-1, change Vel1 to zero and select the SI units. A Calculate produces the new state with a ridiculously large value of A1 (instead of infinity).

Open System Ex. #1: What-if Scenario—Neglect Velocity

Saturated liquid water steadily enters a valve. . . For the what-if study, switch back to State-1, modify it, Calculate, and Super-Calculate.

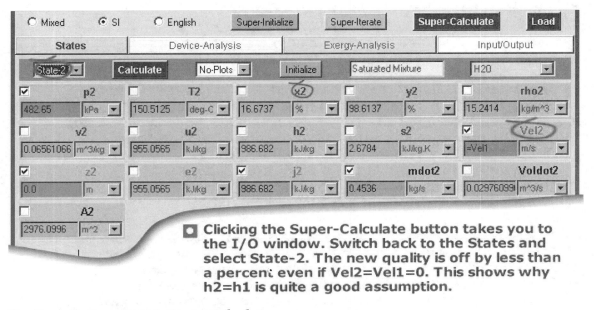

Clicking the Super-Calculate button takes you to the I/O window. Switch back to the States and select State-2. The new quality is off by less than a percent even if Vel2=Vel1=0. This shows why h2=h1 is quite a good assumption.

Open System Ex. #1: Super-Calculate

Saturated liquid water steadily enters a valve. . . The new exit quality is slightly higher. Neglect of ke, therefore, has very little impact on the answer. However, the exit area found, obviously, is not correct.

```
###################################################################
# To solve this problems, copy the following TEST-Code onto the I/O panel of the
# ..Systems.Open.Generic.Steady.SingleFlow.PhaseChanger daemon.
# and click the Load and Super-Calculate buttons.
#--------------------Start of TEST-Code----------------------------------------------------------------------
   States {
   States-0: H2O;
   Given:    { p0= 101.3248 kPa;  T0= 25.0 deg-C; Vel0= 0.0 m/s; z0= 0.0 m;  }

   States-1: H2O;
   Given:    { p1= 2758.0 kPa;  x1= 0.0%; Vel1= 0.0 m/s; z1=0.0 m; mdot1= 0.4536
kg/s;  }

   State-2: H2O;
   Given:    { p2= 482.65 kPa; Vel2= "Vel1" m/s; z2= 0.0 m;  }
   }

   Analysis {
   Device-A: i-State =  State-1; e-State = State-2;
   Given: { Qdot= 0.0 kW; Wdot_O= 0.0 kW; T_B=25.0 deg-C;  }
   }
#--------------------End of TEST-Code------------------------------------------------------------------------
-----
```

■ **Visual Tour: TEST-Code to Regenerate this Solution**

Example: A compressor with a pressure ratio of 5 operates at steady state with R-134a as the working fluid. The refrigerant enters at 0.2 MPa and 0 deg C, with a volumetric flow rate of 0.5 m3/min. The area of the inlet is 8 cm². At the exit, the velocity is 10 m/s and the temperature is 66 deg C. (a) What is the exit area? (b) If the compressor loses heat at a rate of 1 kW, determine the compressor power. (c) Determine the entropy generation rate if the boundary temperature is 100 deg C. (d) If the ambient temperature is 25 deg C, determine the second-law efficiency. (e) How would the answers change if the exit velocity is 20 m/s?

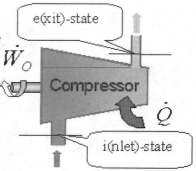

Solution Procedure: Launch the open, steady, single-flow daemon for the phase-change fluid model. Evaluate State-1 and State-2 as the i- and e-states, and State-0 as the dead state. Load the states in the analysis panel, enter the known device variables and Calculate the answers.

■ *Open System Ex. #2: Problem Description*

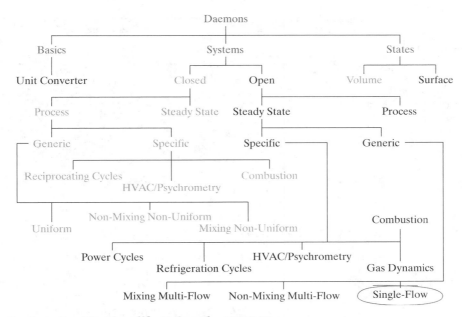

■ *Open System Ex. #2*: **Simplify Using the TEST-Map**

A compressor with a pressure ratio of 5. . . The appropriate simplification for this problem takes us to the. . . Open.Steady.Generic.Single Flow page.

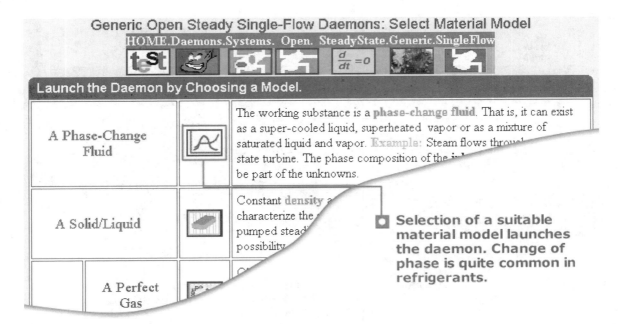

> ☐ **Selection of a suitable material model launches the daemon. Change of phase is quite common in refrigerants.**

■ *Open System Ex. #2*: **Idealization - Pick a Fluid Model**

A compressor with a pressure ratio. . . The phase-change model will produce more accurate answers even though a gas model could also be used for this problem.

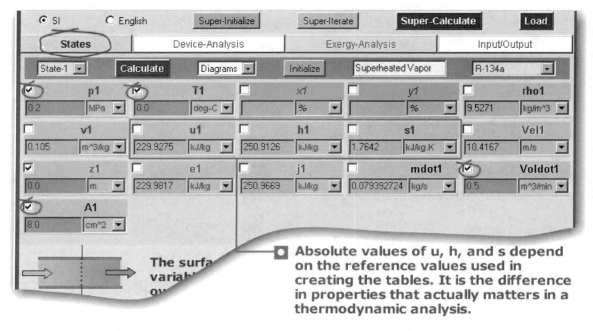

Absolute values of u, h, and s depend on the reference values used in creating the tables. It is the difference in properties that actually matters in a thermodynamic analysis.

Open System Ex. #2: Evaluate the Inlet State (i-State)

A compressor with a pressure ratio... Select the working fluid, enter the known variables and Calculate State-1 as the i-State.

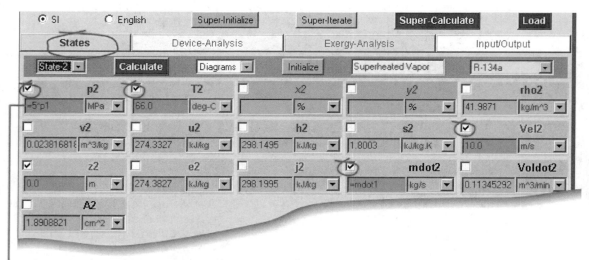

Instead of entering the pressure as 1 MPa, the use of algebraic relation 5*p1 is a better choice. That way, if p1 is changed in a parametric study, the solution can be updated using Super-Calculate without having to alter State-2.

Open System Ex. #2: Evaluate the Exit-State (e-State)

A compressor with a pressure ratio... Enter the known variables using algebraic expressions wherever possible. Calculate State-2 as the e-State.

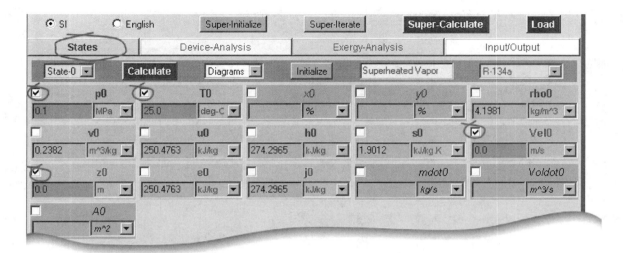

■ *Open System Ex. #2:* **Evaluate the Dead-State**

A compressor with a pressure ratio... Select State-0 as the dead state. Use the ambient pressure and temperature, and Calculate.

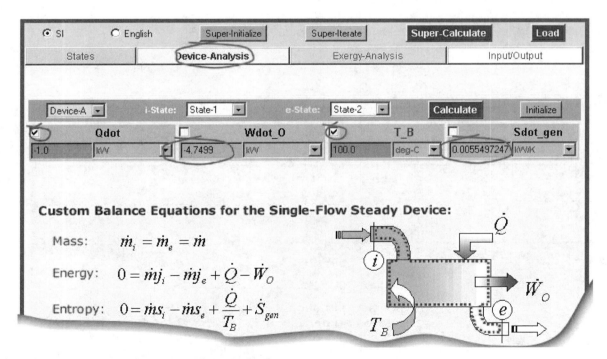

■ *Open System Ex. #2:* **Device Analysis**

A compressor with a pressure ratio... Load the i- and e-states. Enter Qdot and T_B, and Super-Calculate. The power rating and entropy generation rate within the compressor are determined.

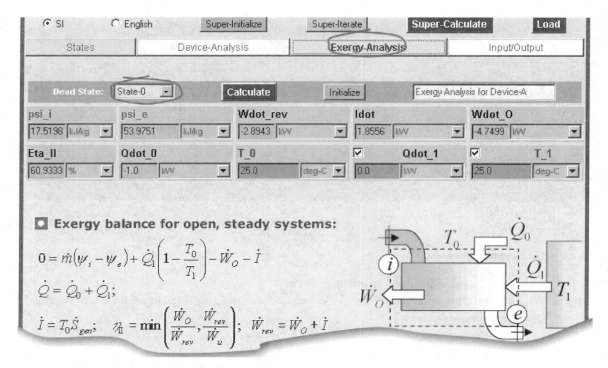

■ *Open System Ex. #2: Exergy Analysis*

A compressor with a pressure ratio. . . Load the dead state and Calculate. All the variables in the exergy balance equation are determined along with the second-law efficiency—eta_II (61%).

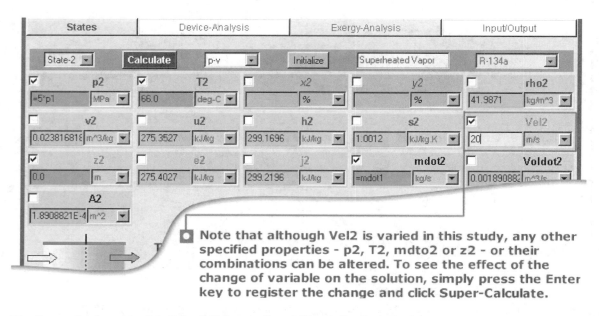

Note that although Vel2 is varied in this study, any other specified properties - p2, T2, mdto2 or z2 - or their combinations can be altered. To see the effect of the change of variable on the solution, simply press the Enter key to register the change and click Super-Calculate.

■ *Open System Ex. #2: What-If Scenario - Effect of Exit Velocity*

A compressor with a pressure ratio. . . Go back to State-2, change Vel2 to the new value, press Enter key (to register the change) and Super-Calculate.

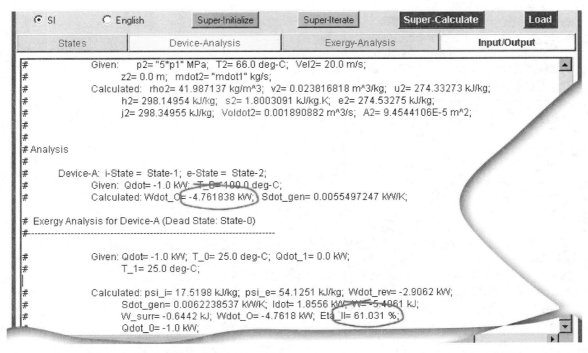

```
# Given:    p2= "5*p1" MPa;  T2= 66.0 deg-C;  Vel2= 20.0 m/s;
#           z2= 0.0 m;  mdot2= "mdot1" kg/s;
# Calculated:  rho2= 41.987137 kg/m^3;  v2= 0.023816818 m^3/kg;  u2= 274.33273 kJ/kg;
#           h2= 298.14954 kJ/kg;  s2= 1.8003091 kJ/kg.K;  e2= 274.53275 kJ/kg;
#           j2= 298.34955 kJ/kg;  Voldot2= 0.001890882 m^3/s;  A2= 9.4544106E-5 m^2;
#
#
# Analysis
#
#     Device-A:  i-State =  State-1;  e-State =  State-2;
#           Given:  Qdot= -1.0 kW;  T_B= 100.0 deg-C;
#           Calculated: Wdot_O= -4.761838 kW;  Sdot_gen= 0.0055497247 kW/K;
#
# Exergy Analysis for Device-A (Dead State: State-0)
#---------------------------------------------------------------------
#
#           Given: Qdot= -1.0 kW;  T_0= 25.0 deg-C;  Qdot_1= 0.0 kW;
#           T_1= 25.0 deg-C;
#
#           Calculated: psi_i= 17.5198 kJ/kg;  psi_e= 54.1251 kJ/kg;  Wdot_rev= -2.9062 kW;
#           Sdot_gen= 0.0062238537 kW/K;  Idot= 1.8556 kW;  W= -5.4061 kJ;
#           W_surr= -0.6442 kJ;  Wdot_O= -4.7618 kW;  Eta_II= 61.031 %;
#           Qdot_0= -1.0 kW;
```

■ **Open System Ex. #2: Super-Calculate**

A compressor with a pressure ratio. . . . All the panels are updated. The detailed solution is displayed on the I/O panel, where the new Wdot_O can be found.

```
#####################################################################
# To regenerate this solution, copy the following TEST-Code onto the I/O panel of the
# ..Open.Steady.Generic.SingleFlow.PhaseChanger
# and click the Load and Super-Calculate buttons.
#-------------------Start of TEST-Code--------------------------------------------
----
 States {
 State-0: R-134a;
 Given:      { p0= 0.1 MPa;  T0= 25.0 deg-C;  Vel0= 0.0 m/s;  z0= 0.0 m;  }

 State-1: R-134a;
 Given:      { p1= 0.2 MPa;  T1= 0.0 deg-C;  z1= 0.0 m;  Voldot1= 0.5 m^3/min;  A1=8.0
cm^2;  }

 State-2: R-134a;
 Given:      { p2= "5*p1" MPa;  T2= 66.0 deg-C;  Vel2=10.0 m/s;  z2= 0.0 m;  mdot2=
"mdot1" kg/s;  }
 }

 Analysis {
  Device-A:  i-State = State-1;  e-State = State-2;
  Given:  {Qdot= -1.0 kW;   T_B= 100.0 deg-C;  }
 }
```

■ **Open System Ex. #2: TEST Code to Regenerate the Visual Solution**

Example: Nitrogen at 500 kPa and 50 deg C enters an adiabatic mixing chamber with a steady flow rate of 1 kg/s, while helium enters at 500 kPa, 500 deg C, and 2 kg/s. There is no pressure drop in the chamber, and KE and PE changes are negligible. (a) Determine the exit composition and temperature, and (b) the rate of entropy generation. (c) How would this answer change if the chamber pressure is 100 kPa instead?

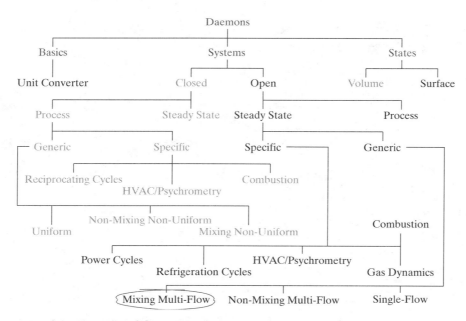

> ⬤ **Solution Procedure: Launch the open, steady, multi-flow, mixing daemon. Choose the ideal gas mixture as the fluid model. Evaluate State-1 and State-2 as the i1- and i2-states, and State-3 as the exit state. Load the states in the analysis panel, enter the known device variables and Super-Calculate the answers.**

■ *Open System Ex. #3:* **Problem Description**

■ *Open System Ex. #3:* **Simplify Using the TEST-Map**

Nitrogen at. . . enters an adiabatic mixing chamber. . . The appropriate simplification for this problem takes us to the ..Open.Steady.Generic.Mixing Multi-Flow page.

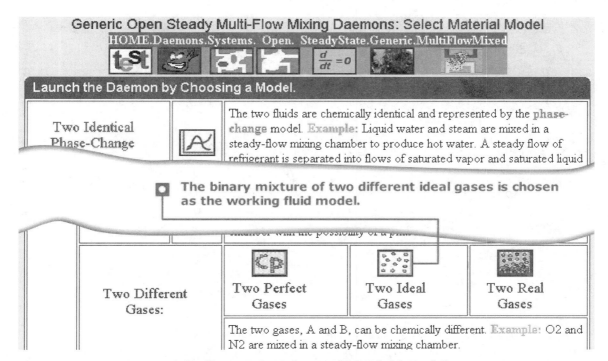

Open System Ex. #3: Idealize – Select the Working Fluid Model

Nitrogen at... enters an adiabatic mixing chamber...The material model we pick here is an ideal gas mixture (with two different gases).

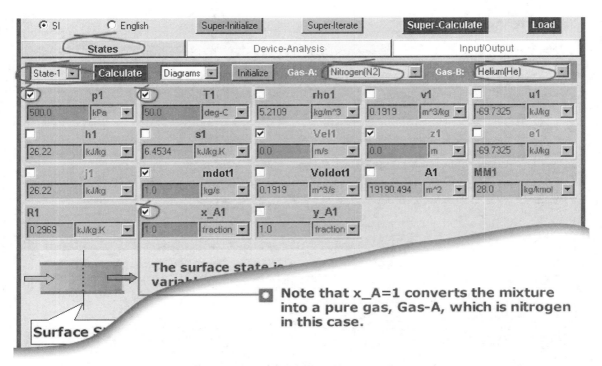

Open System Ex. #3: Evaluate Inlet States (State-1 and State-2)

Nitrogen at... enters an adiabatic mixing chamber... Choose N2 as gas-A and He as gas-B. Enter x_A1 (=1 turns the mixture into pure N2), p1 and T1, and Calculate State-1.

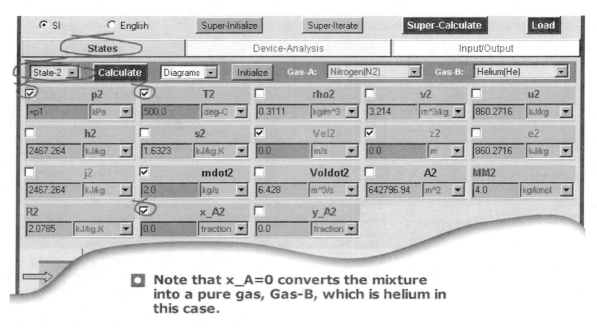

■ **Note that x_A=0 converts the mixture into a pure gas, Gas-B, which is helium in this case.**

■ *Open System Ex. #3*: Evaluate Inlet States (State-1 and State-2)

Nitrogen at. . . enters an adiabatic mixing chamber. . . Enter the mass fraction of N2, x_A2(=0 turns the mixture into pure He), p2 and T2, and Calculate State-2.

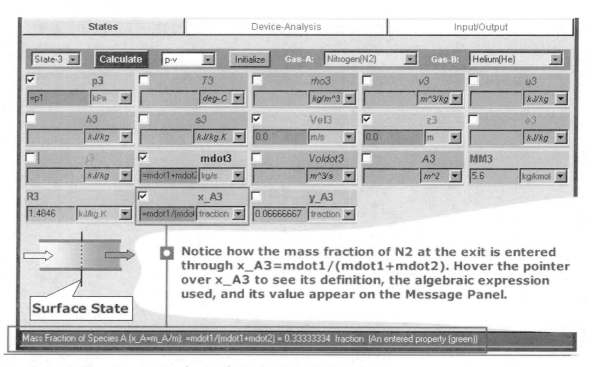

■ **Notice how the mass fraction of N2 at the exit is entered through x_A3=mdot1/(mdot1+mdot2). Hover the pointer over x_A3 to see its definition, the algebraic expression used, and its value appear on the Message Panel.**

Mass Fraction of Species A (x_A=m_A/m): =mdot1/(mdot1+mdot2) = 0.33333334 fraction (An entered property (green))

■ *Open System Ex. #3*: Evaluate the Exit State (State-3)

Nitrogen at. . . enters an adiabatic mixing chamber. . . Enter p3, mass fraction of N2 at the exit (x_A3), and mdot3 as algebraic expressions. Calculate the state.

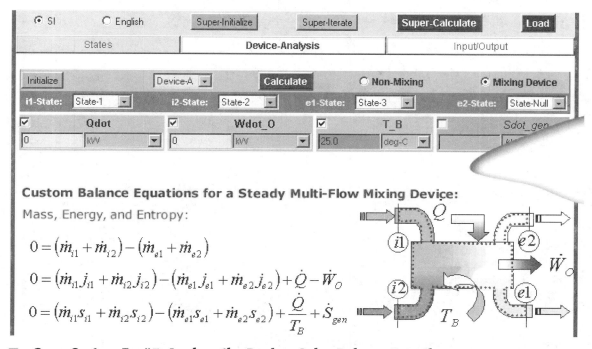

Open System Ex. #3: Analyze the Device—Solve Balance Equations

Nitrogen at... enters an adiabatic mixing chamber... Load the two i-states and State-3 as either e1- or e2-state. Enter the device variables (Qdot and Wdot_O), Calculate, and Super-Calculate.

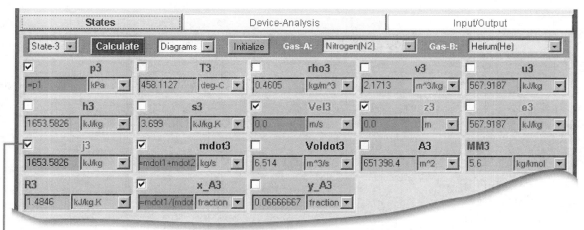

The device analysis produces j3, which is posted back to State-3. Notice the background colors:

   Yellow for inserting input variables.   Green after a variable is read.

   Cyan for calculated variables.   Gray when a variable is posted.

Open System Ex. #3: State-3 Evaluated

Nitrogen at... enters an adiabatic mixing chamber... Now that j3, calculated from the balance equations, is posted, State-3 evaluation is completed by the Super-Calculate operation.

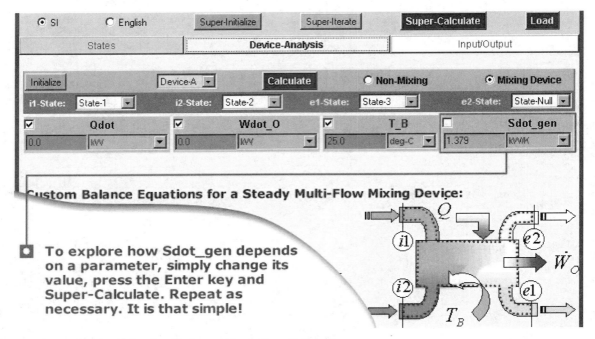

To explore how Sdot_gen depends on a parameter, simply change its value, press the Enter key and Super-Calculate. Repeat as necessary. It is that simple!

■ *Open System Ex. #3*: Entropy Equation Solved

Nitrogen at. . . enters an adiabatic mixing chamber. . . On the Device-Analysis panel, the rate of entropy generation is updated by the Super-Calculate operation after State-3 is found.

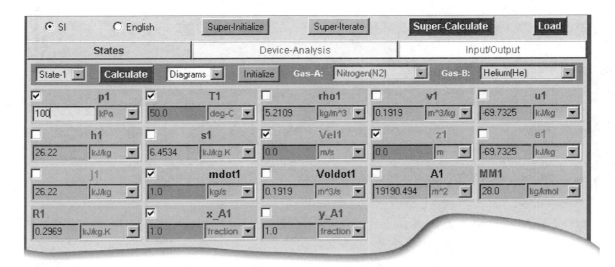

■ *Open System Ex. #3*: Parametric Study—Change p1

Nitrogen at. . . enters an adiabatic mixing chamber. . . On the States panel, change p1 to 100 kPa. Calculate (or simply press the Enter key) and Super-Calculate.

States	Device-Analysis	**Input/Output**

```
#         Given:     p2= "p1" kPa;  T2= 500.0 deg-C;  Vel2= 0.0 m/s;
#                    z2= 0.0 m;  mdot2= 2.0 kg/s;  x_A2= 0.0 fraction;
#
#         Calculated:  rho2= 0.31114027 kg/m^3;  v2= 3.2139845 m^3/kg;  u2= 860.2716 kJ/kg;
#                    h2= 2467.264 kJ/kg;  s2= 1.6322678 kJ/kg.K;  e2= 860.2716 kJ/kg;
#                    j2= 2467.264 kJ/kg;  Voldot2= 6.427969 m^3/s;  A2= 642796.94 m^2;
#                    MM2= 4.0 kg/kmol;  R2= 2.0785 kJ/kg.K;  y_A2= 0.0 fraction;
#
#
#     State-3:  Nitrogen(N2), Helium(He) > IdealGasMixture;
#         Given:     p3= "p1" kPa;  Vel3= 0.0 m/s;  z3= 0.0 m;
#                    mdot3= "mdot1+mdot2" kg/s;  x_A3= "mdot1/(mdot1+mdot2)" fraction;
#         Calculated:  T3= 458.1127 deg-C;  rho3= 0.4605477 kg/m^3;  v3= 2.1713278 m^3/kg;
#                    u3= 567.9187 kJ/kg;  h3= 1653.5826 kJ/kg;  s3= 3.6990016 kJ/kg.K;
#                    e3= 567.9187 kJ/kg;  j3= 1653.5826 kJ/kg;  Voldot3= 6.5139837 m^3/s;
#                    A3= 651398.4 m^2;  MM3= 5.6 kg/kmol;  R3= 1.4846429 kJ/kg.K;
#                    y_A3= 0.06666667 fraction;
#
# Analysis
#
#     Device-A:  i-State = State-1, State-2;  e-State = State-3; Mixing: true
#         Given:  Qdot= 0.0 kW;  Wdot_O= 0.0 kW;  T_B= 25.0 deg-C;
#         Calculated: Sdot_gen= 1.3790221 kW/K;
```

■ *Open System Ex. #3:* **Super-Calculate Updates All Variables**

Nitrogen at… enters an adiabatic mixing chamber. … All panels are updated. The I/O panel displays the detailed report as well as the TEST-Code. The new temperature can be found here.

```
###########################################################################
# To regenerate this solution, copy the following TEST-Code onto the I/O panel of the
# ..Open.Steady.Generic.MultiFlowMixed.IdealGasMixture
# and click the Load and Super-Calculate buttons.
#--------------------Start of TEST-Code-------------------------------------------------------
----
 States {
 State-1:  Nitrogen(N2), Helium(He);
 Given:     { p1= 500.0 kPa;  T1= 50.0 deg-C;  Vel1= 0.0 m/s;  z1= 0.0 m;  mdot1= 1.0
kg/s;  x_A1= 1.0 fraction;  }

 State-2:  Nitrogen(N2), Helium(He);
 Given:     { p2= "p1" kPa;  T2= 500.0 deg-C;  Vel2= 0.0 m/s;  z2= 0.0 m;  mdot2= 2.0
kg/s;  x_A2= 0.0 fraction;  }

 State-3:  Nitrogen(N2), Helium(He);
 Given:     { p3= "p1" kPa;  Vel3= 0.0 m/s;  z3= 0.0 m;  mdot3= "mdot1+mdot2" kg/s;
x_A3= "mdot1/(modt1+mdot2)" fraction;  }
 }

 Analysis {
  Device-A:  i-State = State-1, State-2; e-State = State-3; Mixing: true
  Given:  {Qdot= 0.0 kW; Wdot_O= 0.0 kW; T_B= 25.0 deg-C; }
 }
```

■ *Open System Ex. #3:* **TEST-Code to Regenerate the Visual Solution**

Example: Water enters a radiator with a flow rate of 2 kg/s at 0.5 MPa and 80 deg C and leaves at 40 deg C at the same pressure. Air enters the radiator at 0.1 MPa and 20 deg C and leaves 10 deg C below the exit temperature of water at the same pressure. Neglecting any heat losses, determine (a) the mass flow rate ratio (b) the rate of entropy generation if the ambient temperature is 20 deg C. How would these answers change if the mass flow rate of water doubles?

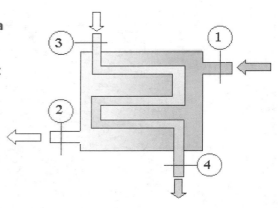

> ◻ Solution Procedure: Launch the open, steady, non-mixing, multiflow daemon with phase-change fluid as the working fluid model. Evaluate the inlet and exit states, load the states into the analysis panel, enter the known device variables, and Super-Calculate the answers.

■ *Open System Ex. #4*: **Problem Description**

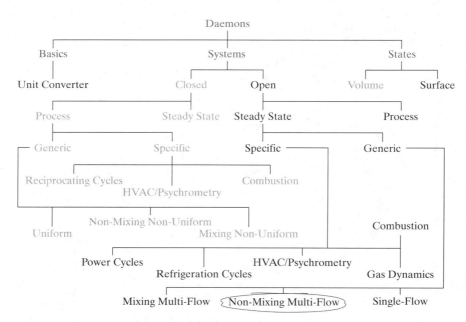

■ *Open System Ex. #4*: **Simplify Using the TEST-Map**

Water enters a radiator with a flow rate of... The appropriate simplification leads us to the ..Open.Steady.Generic.Non-Mixing-multi-flow page.

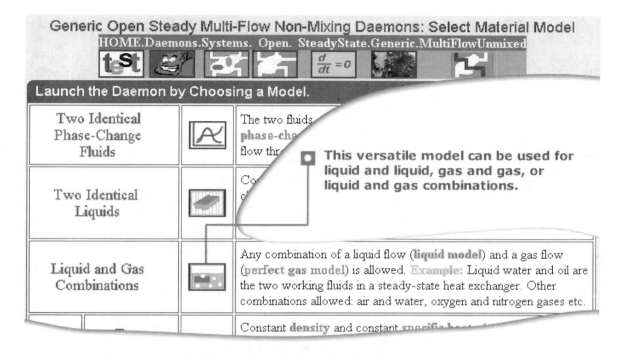

Open System Ex. #4: Idealize – Select Material Model

Water enters a radiator with a flow rate of. . . The material models we pick here are liquid an perfect gas for the two fluids.

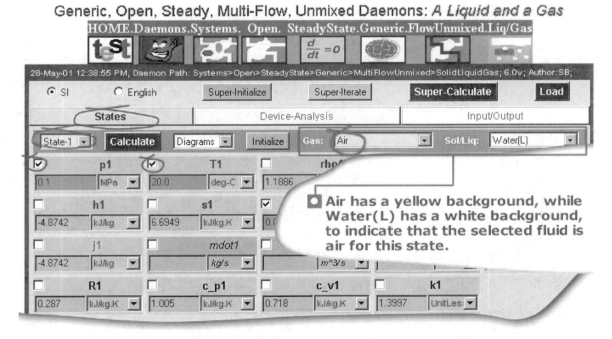

Open System Ex. #4: Evaluate the Air Inlet State (State-1)

Water enters a radiator with a flow rate of . . . Choose Air as the working fluid for the i1-state (State-1), enter the known variables and Calculate. The mass flow rate is the desired unknown.

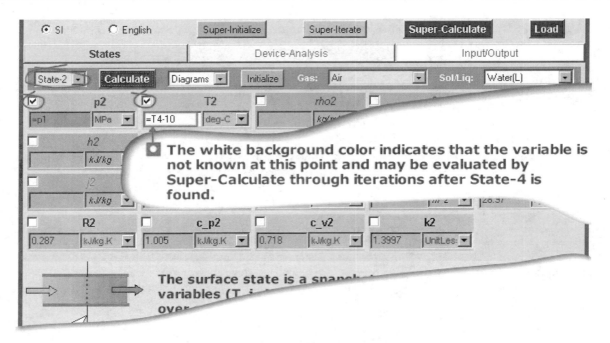

The white background color indicates that the variable is not known at this point and may be evaluated by Super-Calculate through iterations after State-4 is found.

- ■ **Open System Ex. #4: Evaluate the Air Exit State (State-2)**

Water enters a radiator with a flow rate of. . . Evaluate State-2, the exit state for air. Note that T2 is expressed in terms of T4, an unknown at this point.

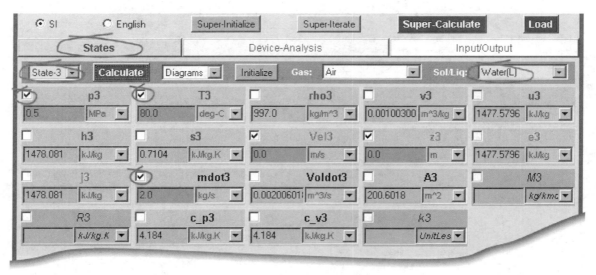

■ Notice that selecting a liquid from the Sol/Liq menu deactivates some of the variables, such as M, R, and k, which are activated when a gas is chosen.

- ■ **Open System Ex. #4: Evaluate the Water Inlet State (State-3)**

Water enters a radiator with a flow rate of. . . Evaluate State-3, the inlet state for liquid water. Select Water(L) as the working fluid and then enter the known variables.

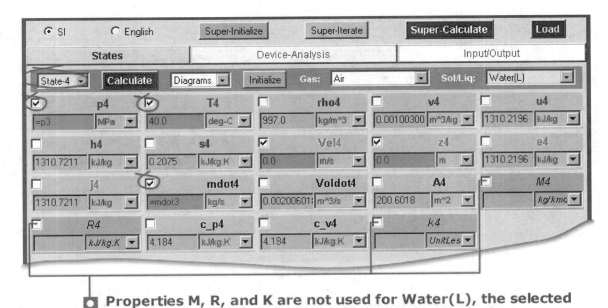

> ◻ Properties M, R, and K are not used for Water(L), the selected fluid for State-4. These are enabled when a gas is chosen.

■ **Open System Ex. #4: Evaluate the Water Exit State (State-4)**

Water enters a radiator with a flow rate of. . .

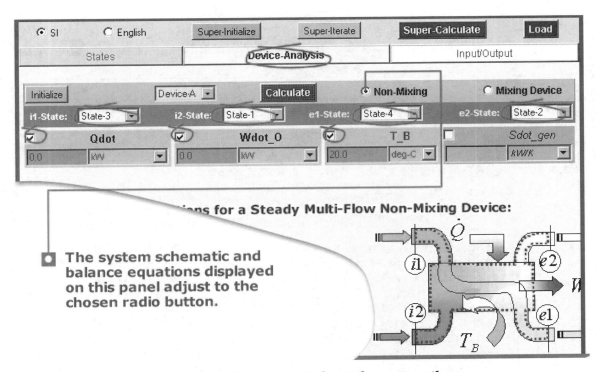

> ◻ The system schematic and balance equations displayed on this panel adjust to the chosen radio button.

■ **Open System Ex. #4: Analyze the Device—Solve Balance Equations**

Water enters a radiator with a flow rate of. . . On the Analysis panel, load the inlet and exit states, make sure that the Non-Mixing button is chosen, enter known device variables, Calculate and Super-Calculate. Note that the balance equations and the device schematic adjust to the toggling between the Mixing and the Non-Mixing buttons.

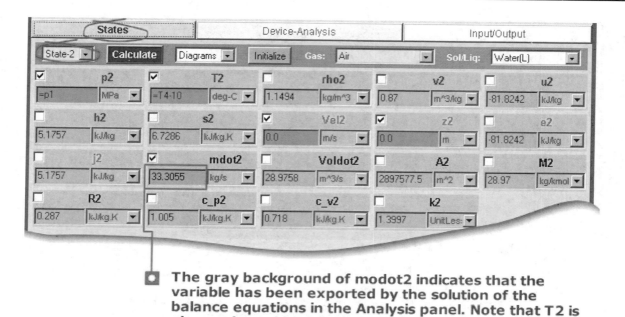

The gray background of modot2 indicates that the variable has been exported by the solution of the balance equations in the Analysis panel. Note that T2 is also evaluated by Super-Calculate.

Open System Ex. #4. State 2 Determined

Water enters a radiator with a flow rate of. . . The Super-Calculate command iterates between the States and Analysis panels to produce the complete solution. All desired variables are not known.

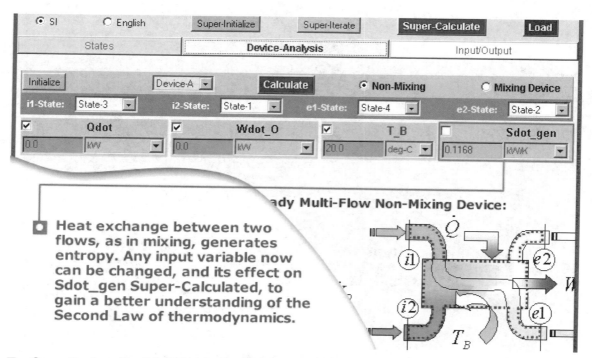

Heat exchange between two flows, as in mixing, generates entropy. Any input variable now can be changed, and its effect on Sdot_gen Super-Calculated, to gain a better understanding of the Second Law of thermodynamics.

Open System Ex. #4: Entropy Generation Rate

Water enters a radiator with a flow rate of. . . The Super-Calculate command updates all the panels, including the Analysis panel, where Sdot_gen can be found.

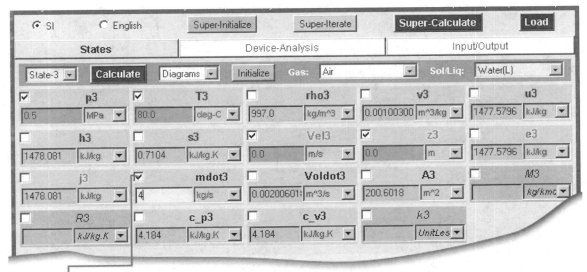

A change in parameter is effective only after you press the Enter key or the Calculate button. Otherwise, Super-Calculate will use the original value for updating all calculations.

■ *Open System Ex. #4:* **Parametric Study—Change mdot3**

Water enters a radiator with a flow rate of. . . Like mdot3 in this example, any input variable (the ones with their boxes checked) can be changed to see its effect on the solution.

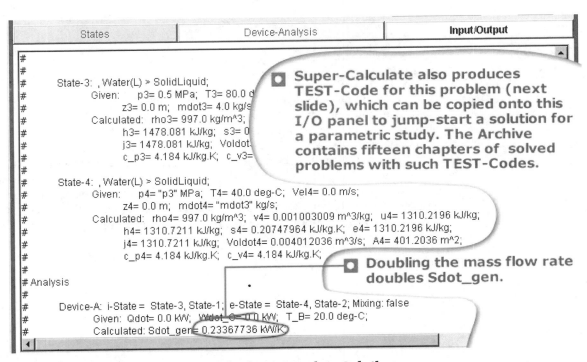

Super-Calculate also produces TEST-Code for this problem (next slide), which can be copied onto this I/O panel to jump-start a solution for a parametric study. The Archive contains fifteen chapters of solved problems with such TEST-Codes.

Doubling the mass flow rate doubles Sdot_gen.

■ *Open System Ex. #4:* **Super-Calculate to Update Solutions**

Water enters a radiator with a flow rate of. . . Click Super-Calculate to update all answers and produce a detailed solution report on the I/O panel.

```
#####################################################################
# To regenerate this solution, copy the following TEST-Code onto the I/O panel of the
# ..Open.Steady.Generic.MultiFlowNonMixing.LiqGas
# and click the Load and Super-Calculate buttons.
#-------------------Start of TEST-Code--------------------------------------------------------
----
States {
 State-1:  Air;
 Given:        { p1= 0.1 MPa;  T1= 20.0 deg-C;  Vel1= 0.0 m/s;  z1= 0.0 m; }

 State-2:  Air;
 Given:        { p2= "p1" MPa;  T2= "T4-10" deg-C;  Vel2= 0.0 m/s;  z2= 0.0 m;  }

 State-3:  ,Water(L);
 Given:        { p3= 0.5 MPa;  T3= 80.0 deg-C;  Vel3= 0.0 m/s;  z3= 0.0 m;  mdot3= 4.0
kg/s;  }

 State-4:  ,Water(L);
 Given:        { p4= "p3" MPa;  T4= 40.0 deg-C;  Vel4= 0.0 m/s;  z4= 0.0 m;  mdot4=
"mdot3" kg/s;  }
}

Analysis {
 Device-A:  i-State = State-3, State-1; e-State = State-4, State-2; Mixing: false
 Given:  { Qdot= 0.0 kW; Wdot_O= 0.0 kW; T_B= 25.0 deg-C;  }
}
```

■ *Open System Ex. #4*: **TEST-Code to Regenerate the Visual Solution**

Open Process Daemons

A clear beginning and a finish mark a process as an open system responds to mass, heat, and work interactions with its surroundings. The states that anchor an open process are the b- and f-states of the system, and the i- and e-states at the inlet and exit ports.

The process analysis panel for Open Process daemons is a combination of the analysis panels of the Closed Process and Open Steady daemons. Anchor states calculated on the state panel are imported, known process variables are entered, and the balance equations are solved by Calculate on the process panel. The Super-Calculate iterates between the two panels coupling the balance equations to the state equations.

Example: A completely evacuated, rigid, well-insulated cylinder develops a leak and gets quickly filled with the surrounding air, which is at 100 kPa and 25 deg C. Determine the temperature inside the cylinder after it comes to a pressure equilibrium with the surroundings.

What-If Scenario: (a) If the volume of the tank were 100 gallons, determine the entropy generation during the process. **(b)** How would the answers change if the cylinder were charged with helium instead of air?

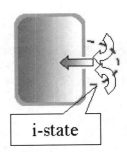

i-state

🔲 **Solution Procedure: Simplify the problem as an open system undergoing a process. Select ideal gas as the material model. Evaluate the begin, finish, and inlet states, load the states in the analysis panel, enter the known process variables, and Super-Calculate the answers.**

■ *Open System Ex. #5.* **Problem Description**

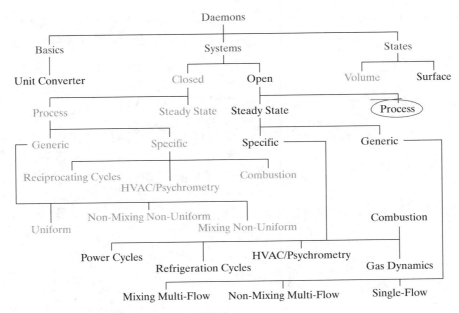

■ *Open System Ex. #5:* **Simplify and Idealize**

A completely evacuated, well-insulated. . . Use systematic navigation or the TEST-map to get to the right daemon page, . . . Open.Process, for this problem. Choose the ideal gas model to launch the daemon.

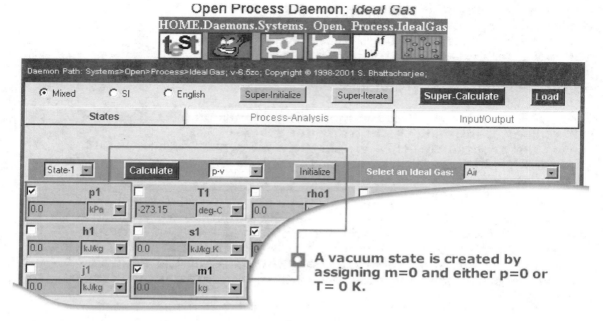

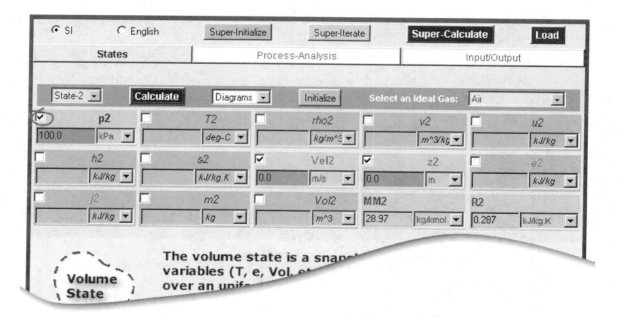

■ *Open System Ex. #5*: Evaluate the Begin State

A completely evacuated well-insulated...To represent vacuum, enter m = 0. Note that even if volume of the cylinder is given, it should not be entered here. The volume of the rigid cylinder this special case, is not the same as the volume of air.

■ *Open System Ex. #5*: Evaluate the Finish State

A completely evacuated, well-insulated... With only the final pressure known for state-2, the designated f-state, the state evaluation is incomplete at this point.

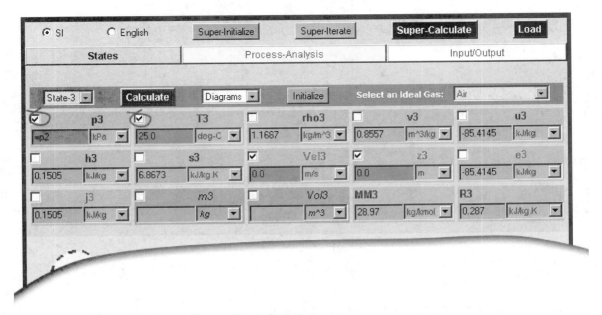

Open System Ex. #5: Evaluate the Inlet State

A completely evacuated, well-insulated... By taking the control surface for the i-state outside and away from the hole, the velocity of air can be neglected. Moreover, at the hole, the pressure and temperature may not be the same as the corresponding ambient values.

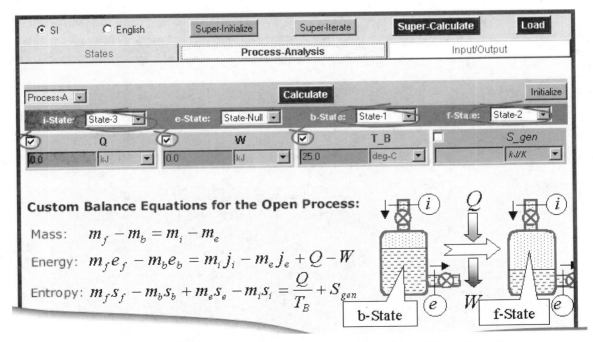

Open System Ex. #5: Analyze the Open Process—Solve the Balance Equations

A completely evacuated, well-insulated... Carefully load the appropriate inlet and exit states. Enter Q (insulated) and W (rigid), and Super-Calculate.

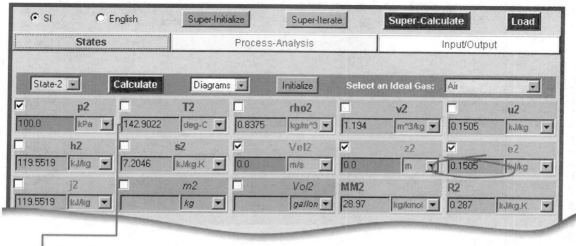

> The temperature after the tank fills up is much above the ambient temperature.

■ *Open System Ex. #5*: Final Temperature Determined

A completely evacuated, well-insulated. . . The solution of the balance equations produces e2, which is posted back to state-2 (note the background color). The final temperature, T2, is determined when Super-Calculate updates all states.

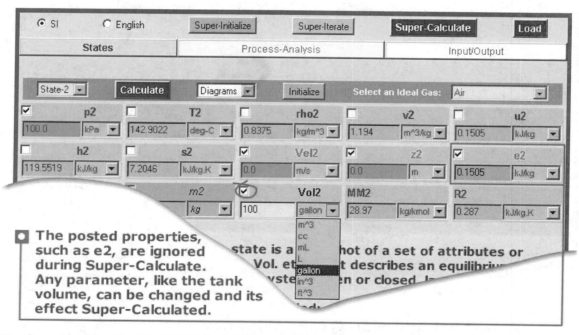

> The posted properties, such as e2, are ignored during Super-Calculate. Any parameter, like the tank volume, can be changed and its effect Super-Calculated.

■ *Open System Ex. #5*: Additional Information—Tank Volume

A completely evacuated, well-insulated. . . Enter the tank volume and Super-Calculate.

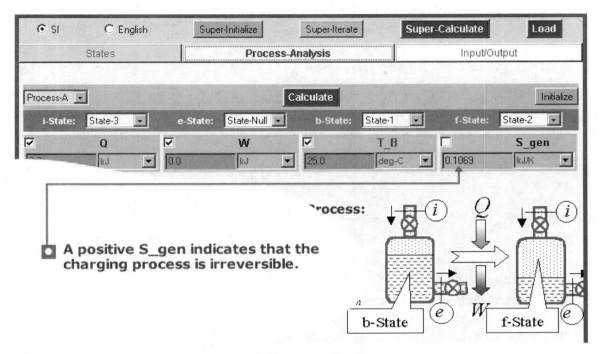

A positive S_gen indicates that the charging process is irreversible.

■ *Open System Ex. #5*: **Super-Calculate to Update Solution**

A completely evacuated, well-insulated. . . Super-Calculate produces the entropy generation during the charging process.

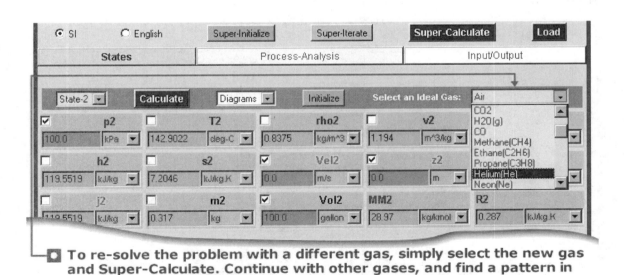

To re-solve the problem with a different gas, simply select the new gas and Super-Calculate. Continue with other gases, and find a pattern in terms of irreversibility and final temperature exhibited by ideal gases.

■ *Open System Ex. #5*: **Parametric Study–Change Working Fluid**

A completely evacuated, well-insulated. . . For the parametric study, select helium in the States panel.

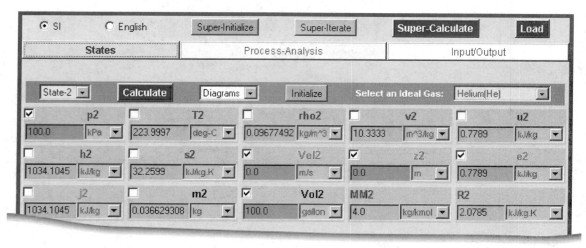

🔲 The final temperature is much higher
when helium, a lighter gas, fills the tank.

■ *Open System Ex. #5*: Super-Calculate

A completely evacuated, well-insulated... Super-Calculate produces a much higher final temperature, 224 deg-C, as opposed to 143 deg-C for air. Why?

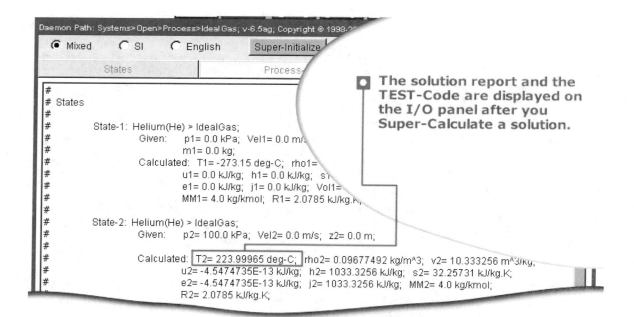

🔲 The solution report and the
TEST-Code are displayed on
the I/O panel after you
Super-Calculate a solution.

■ *Open System Ex. #5*: Solution Report on the I/O Panel

A completely evacuated, well-insulated... Super-Calculate changes the focus to the Input/Output panel, where the TEST-Code, property table and the complete solution report are displayed.

```
###########################################################################
# To regenerate this solution, copy the following TEST-Code onto the I/O panel of the
# ..Open.Process.IdealGas daemon
# and click the Load and Super-Calculate buttons.
#--------------------Start of TEST-Code-------------------------------------------------------------
----
 States {
 State-1:  Air;
 Given:      { Vel1= 0.0 m/s;  z1= 0.0 m;  m1= 0.0 kg;  }

 State-2:  Air;
 Given:      { p2= 100.0 kPa;  Vel2= 0.0 m/s;  z2= 0.0 m;  }

 State-3:  Air;
 Given:      { p3= 100.0 kPa;  T3= 25.0 deg-C;  Vel3= 0.0 m/s;  z3= 0.0 m;  }
 }

Analysis {
 Process-A:  ie-State = State-3, State-Null;  bf-State = State-1, State-2;
 Given:  { Q= 0.0 kJ;   W= 0.0 kJ;   T_B= 25.0 deg-C;  }
 }
```

■ *Open System Ex. #5*: **TEST-Code to Regenerate the Visual Solution**

States Advanced Topics

Except for simple models such as the perfect gas model or the solid/liquid model, where closed-form analytical expressions are available for evaluating thermodynamic properties, state evaluation of most other materials generally involve tables, charts, or software like TEST. The science behind creating those thermodynamic tables is the subject matter of this chapter. Instead of recreating a table using the thermodynamic relations, such as the Maxwell's equation, T - dS relation, etc., a reverse approach is used. Use of the state relations allows us to test a database, and derive new properties that are not supplied as part of the database. The first example shows how a partial derivative can be evaluated using a State daemon. Advanced state relations necessary to evaluate the state of mixtures are also covered in this chapter. Mixture problems, it should be mentioned, can also be found in the Chapter 2, 6 and 8.

In evaluating partial derivatives, the round off error should be kept in mind. When two states are made to approach each other, a point is reached beyond which the round-off errors will become significant, and an expected limiting value may not materialize.

Example: Determine the Joule–Thompson coefficient of saturated ammonia vapor at 100 kPa.

How would the answer change if ammonia is superheated to 30 deg C (at 100 kPa)?

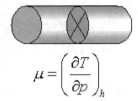

$$\mu = \left(\frac{\partial T}{\partial p}\right)_h$$

◻ **Solution Procedure: Simplify the problem as a state problem with phase-change fluid model. Evaluate the given state and a neighboring state holding h=constant. The coefficient is obtained by evaluating delta_T/delta_p.**

■ *State Relations Ex. #1*: **Problem Description**

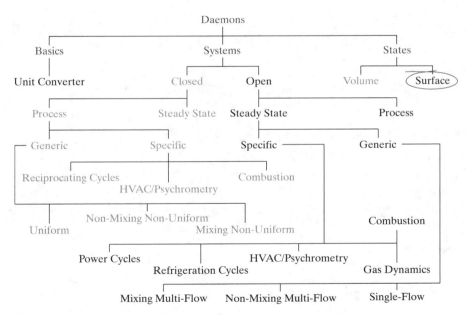

■ *State Relations Ex. #1*: **Simplify the Problem**

Determine the Joule–Thompson... One can treat this problem as a state evaluation problem. Select Daemons.States.Surface page and choose the phase-change model.

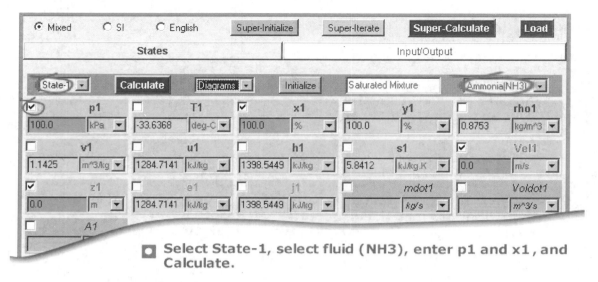

□ Select State-1, select fluid (NH3), enter p1 and x1, and Calculate.

■ *State Relations Ex. #1*: **Evaluate State-1**

Determine the Joule–Thompson. . . Saturated vapor means that the quality is 100%.

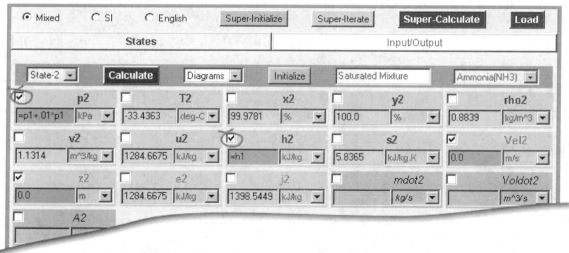

□ Select State-2, enter p2 (=p1+0.01*p1) and h2 (=h1), and Calculate.

■ *State Relations Ex. #1*: **Evaluate a Neighborhood State. State-2**

Determine the Joule–Thompson. . . A neighboring state to State-1 is obtained by perturbing p1 (by 1%) and holding h constant (h2 = h1).

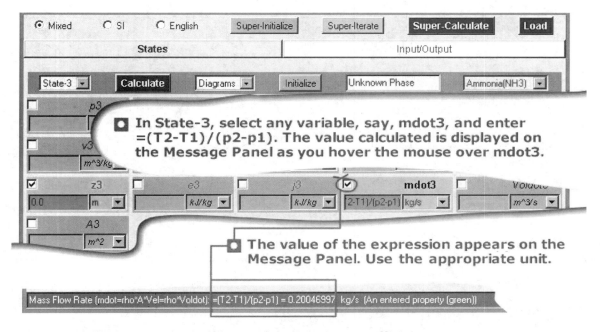

■ *State Relations Ex. #1*: Evaluate Joule–Thompson Coefficient

Determine the Joule–Thompson. . . . Select State-3. Use mdot3 as a dummy variable to evaluate the partial derivative of T with respect to p. Obviously, the unit should be deg-C/kPa.

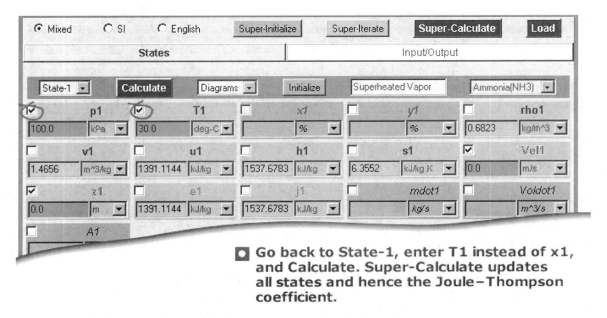

■ *State Relations Ex. #1*: Change a Parameter–Temperature

Determine the Joule–Thompson. . . . Go back to State-1, enter T1 instead of x1, and Calculate. Super-Calculate to update all calculations.

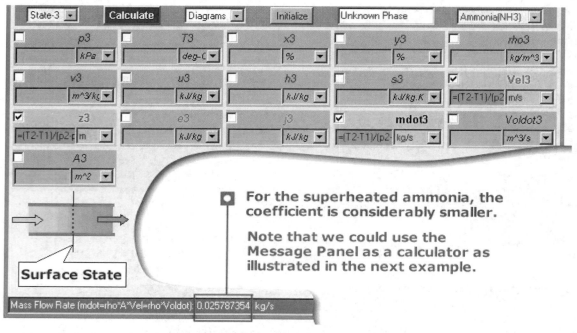

For the superheated ammonia, the coefficient is considerably smaller.

Note that we could use the Message Panel as a calculator as illustrated in the next example.

Surface State

Mass Flow Rate (mdot=rho*A*Vel=rho*Voldot): 0.025787354 kg/s

■ *State Relations Ex. #1*: **New Coefficient**

Determine the Joule–Thompson. . . The coefficient is much smaller now because super heated ammonia behaves more like an ideal gas, for which the JT coefficient is zero.

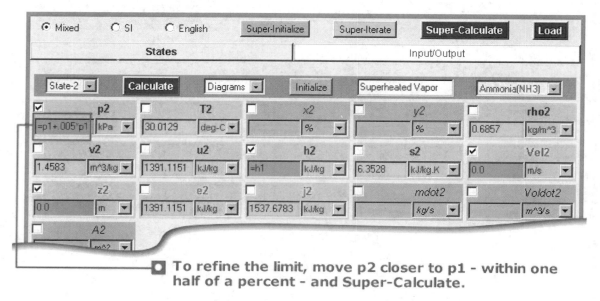

■ To refine the limit, move p2 closer to p1 - within one half of a percent - and Super-Calculate.

■ *State Relations Ex. #1*: **Sensitivity to Delta-p**

Determine the Joule–Thompson. . . How accurate is the partial derivative? To check its accuracy, reduce the perturbation in p (from 1% to .05%) and Super-Calculate.

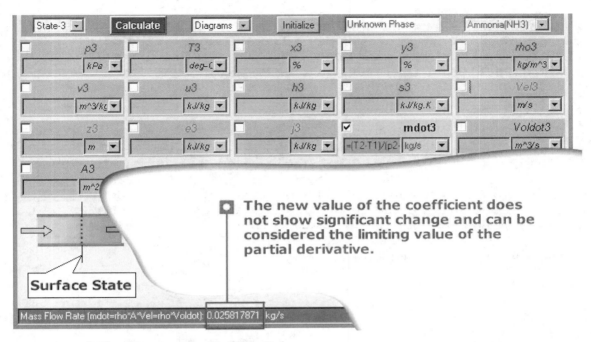

The new value of the coefficient does not show significant change and can be considered the limiting value of the partial derivative.

Surface State

Mass Flow Rate (mdot=rho*A*Vel=rho*Voldot): 0.025817871 kg/s

■ *State Relations Ex. #1:* **The Limiting Value**

Determine the Joule–Thompson. . . The new value of the partial derivative is hardly any different, indicating that it has been correctly evaluated.

```
######################################################################
# To regenerate this solution, copy the following TEST-Code onto the I/O panel of the
# ...States.Surface.PhaseChange daemon
# and click the Load and Super-Calculate buttons.
#--------------------Start of TEST-Code--------------------------------
----

States {
State-1:  H2O;
Given:      {P1= 100.0 kPa;  x1= 100.0 %;  Vel1= 0.0 m/s;  z1= 0.0 m;  }

State-2: H2O;
Given:       { p2 = "p1+.01*p1" kPa;  h2 = "h1" kJ/kg;  Vel2 = 0.0 m/s;  z2 = 0.0 m;  }

State-3: H2O;
Given:       { Vel3 = 0.0 m/s;  z3 = 0.0 m;  mdot3 = "(T2-T1)/(p2-p1)" kg/s;  }
}
```

■ *State Relations Ex. #1:* **TEST-Code to Regenerate the Visual Solution**

Example: A 50–50 mixture of N2 and O2 by mass is heated from 250 K to 350 K at a constant pressure of 10 MPa. Determine the change in enthalpy and entropy assuming the mixture to follow the real gas mixture model using Kay's rule. How would the answers change if the ideal gas mixture model were used instead?

○ **Solution Procedure: In TEST all state related problems are treated in a uniform manner. Launch the Volume State daemon for the ideal gas mixture model. Evaluate the two states and copy the TEST-Code into the real gas mixture State daemon. Load and Super-Calculate the desired states.**

■ *States Relations Ex. #2:* **Problem Description**

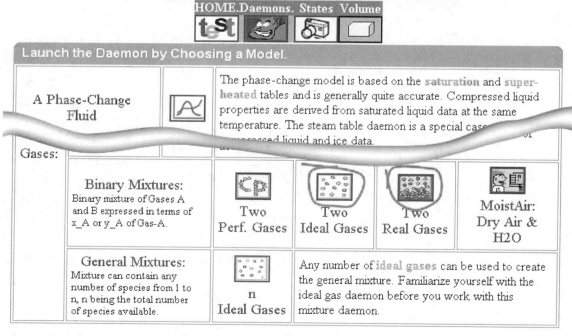

The Volume State Daemons: Select Material Model

HOME.Daemons. States Volume

Launch the Daemon by Choosing a Model.

| A Phase-Change Fluid | | The phase-change model is based on the saturation and super-heated tables and is generally quite accurate. Compressed liquid properties are derived from saturated liquid data at the same temperature. The steam table daemon is a special case ~~compressed liquid and ice data.~~ |

Gases:

| | Binary Mixtures: Binary mixture of Gases A and B expressed in terms of x_A or y_A of Gas-A. | Two Perf. Gases | Two Ideal Gases | Two Real Gases | MoistAir: Dry Air & H2O |
| | General Mixtures: Mixture can contain any number of species from 1 to n, n being the total number of species available. | n Ideal Gases | Any number of ideal gases can be used to create the general mixture. Familiarize yourself with the ideal gas daemon before you work with this mixture daemon. | | |

■ *State Relations Ex. #2:* **Simplify and Launch Daemon**

A 50–50 mixture of nitrogen and oxygen. . . Use the Map to launch the mixture daemons with real gas and ideal gas models.

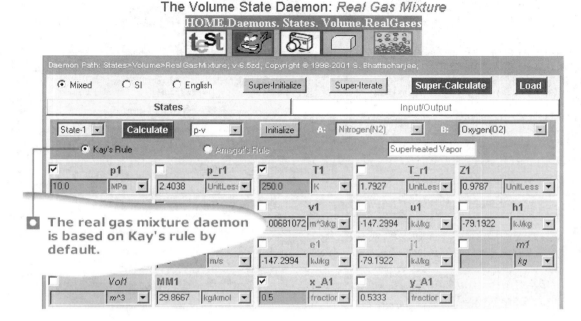

State Relations Ex. #2: Evaluate State-1

A 50–50 mixture of nitrogen and oxygen... Enter the mass fraction of nitrogen, x__A1, p1, and Calculate. Note that Kay's rule is applied by default for a real gas mixture.

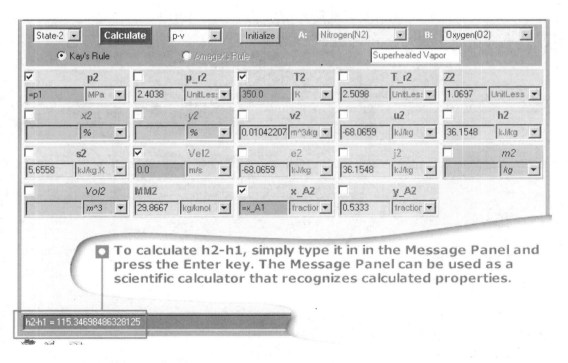

State Relations Ex. #2: Evaluate State-2

A 50–50 mixture of nitrogen and oxygen... Note the use of the built-in calculator, which can used like a scientific calculator (including use of trigonometric functions). Any calculated property is also recognized and symbolic operations, such as h2-h1, s2-s1, etc., are allowed.

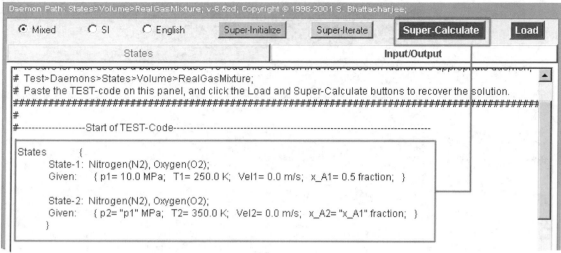

Daemon Path: States>Volume>RealGasMixture; v-6.5zd; Copyright © 1998-2001 S. Bhattacharjee;

```
#  Test>Daemons>States>Volume>RealGasMixture;
# Paste the TEST-code on this panel, and click the Load and Super-Calculate buttons to recover the solution.
####################################################################################
#
#------------------Start of TEST-Code--------------------------------------------------

States      {
        State-1:  Nitrogen(N2), Oxygen(O2);
        Given:     { p1= 10.0 MPa;  T1= 250.0 K;  Vel1= 0.0 m/s;  x_A1= 0.5 fraction;  }

        State-2:  Nitrogen(N2), Oxygen(O2);
        Given:     { p2= "p1" MPa;  T2= 350.0 K;  Vel2= 0.0 m/s;  x_A2= "x_A1" fraction;  }
        }

-----------------End of TEST-Code-----------------------------------------------------
```

☐ **Copy this TEST-Code to the I/O panel of the ideal gas mixture daemon.**

■ *State Relations Ex. #2:* **Generate TEST-Code**

A 50–50 mixture of nitrogen and oxygen. . . Super-Calculate to generate the TEST-Code. To copy the code, select by dragging the mouse, use Ctrl-C, open the ideal gas mixture State daemon, click anywhere on the I/O panel, and then use Ctrl-V.

The Volume State Daemon: *Ideal Gas Mixture*

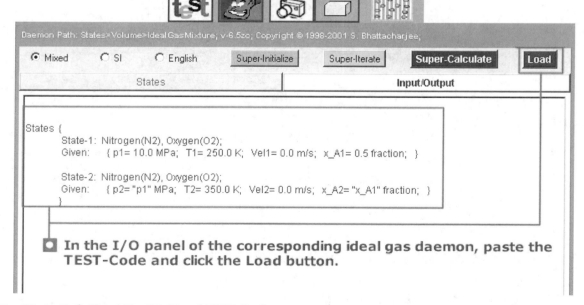

HOME.Daemons. States. Volume.IdealGases

Daemon Path: States>Volume>IdealGasMixture; v-6.5zc; Copyright © 1998-2001 S. Bhattacharjee;

```
States {
        State-1:  Nitrogen(N2), Oxygen(O2);
        Given:     { p1= 10.0 MPa;  T1= 250.0 K;  Vel1= 0.0 m/s;  x_A1= 0.5 fraction;  }

        State-2:  Nitrogen(N2), Oxygen(O2);
        Given:     { p2= "p1" MPa;  T2= 350.0 K;  Vel2= 0.0 m/s;  x_A2= "x_A1" fraction;  }
        }
```

☐ **In the I/O panel of the corresponding ideal gas daemon, paste the TEST-Code and click the Load button.**

■ *State Relations Ex. #2:* **Load TEST-Code**

A 50–50 mixture of nitrogen and oxygen. . . Simply click the Load button. A pop-up window will display the loading messages.

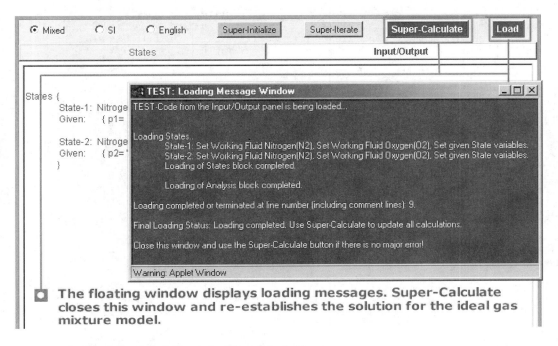

The floating window displays loading messages. Super-Calculate closes this window and re-establishes the solution for the ideal gas mixture model.

■ *State Relations Ex. #2:* Generate Visual Solution

A 50–50 mixture of nitrogen and oxygen. . . . Super-Calculate to generate the complete solution the ideal gas mixture model. The pop-up window can be closed manually or automatically by use of the Super-Calculate button.

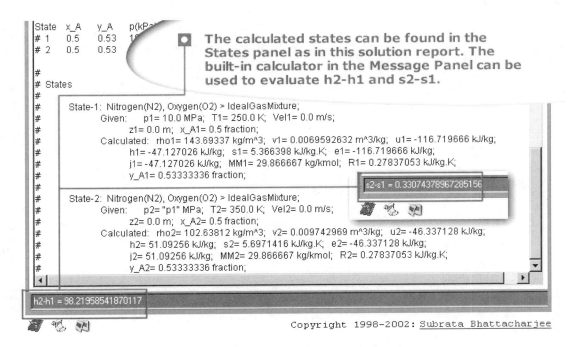

The calculated states can be found in the States panel as in this solution report. The built-in calculator in the Message Panel can be used to evaluate h2-h1 and s2-s1.

■ *State Relations Ex. #2:* Solution for the Ideal Gas Mixture Model

A 50–50 mixture of nitrogen and oxygen. . . . Use the built-in calculator to evaluate h2-h1. (Simply type in the expressions on the Message Panel and press the Enter key). At the given pressure, which is quite high compared to the critical pressure of the mixture, the ideal gas mixture model is, obviously, quite inaccurate.

```
###########################################################################
# To regenerate this solution, copy the following TEST-Code onto the I/O panel of the
# ..States. Volume.RealGasMixture or IdealGasMixture daemon
# and click the Load and Super-Calculate buttons.
#--------------------Start of TEST-Code-------------------------------------------------------------
----

States {
State-1:  Nitrogen(N2), Oxygen(O2);
Given:      { p1 = 10.0 MPa;   T1 = 250.0 K;   Vel1 = 0.0 m/s;   z1 = 0.0 m;  x_A1 = 0.5
fraction;  }

State-2: Nitrogen(N2), Oxygen(O2);
Given:      { p2 = "p1" Mpa;   T2 = 350.0 K;  Vel2 = 0.0 m/s  z2 = 0.0 m;  x_A2 = 0.5
fraction;  }
}
```

■ *States Relations Ex. #2*: **TEST-Code to Generate the Visual Solution**

Closed-Process Cycles

To emphasize the fact that reciprocating air-standard cycles, such as the Otto and Diesel cycles, are based on a succession of closed processes they are called closed-process cycles in TEST, and the corresponding daemons are named Closed-Process daemons. Although only a single example is given here, the Tutorial and Chapter 8 in the Archive contain a number of examples with different types of reciprocating cycles.

The key to understanding these daemons is to work with the Closed-Process daemons covered in Chapter 3. A **Cycle** panel is added as an additional layer to the Process-Analysis panel. After each process of the cycle is analyzed, the cycle variables are automatically calculated. The Cycle panel checks to make sure that the cycle is complete by checking that the finish state of the last process is the begin state of the first process.

Example: An air-standard cycle is executed
in a closed system with 1 kg of air, and it
consists of the following three processes: (a)
isentropic compression from 100 kPa, 27
deg C, to 700 kPa; (b) p=constant heat
addition to initial specific volume; (c)
v=constant heat rejection to initial state.

(i) Calculate the maximum temperature and
the efficiency of the cycle. (ii) Show the cycle
on a T–s and a p–v diagram. Assume variable
cp. How would the answers change if the
pressure after the compression stroke were
changed to 1000 kPa?

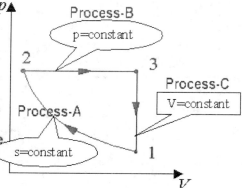

> ☐ Solution Procedure: Simplify the problem as a specific closed cycle.
> Use the ideal gas model. Evaluate the begin and finish states of all the
> three component processes, analyze each process, and finally, get all
> the cycle-related answers on the cycle panel.

■ *IC-Engines Ex. #1*: **Problem Description**

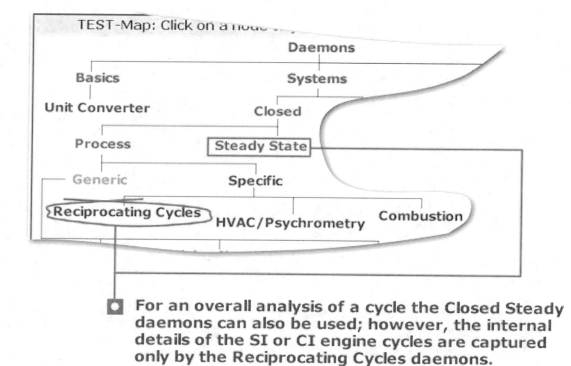

☐ **For an overall analysis of a cycle the Closed Steady
daemons can also be used; however, the internal
details of the SI or CI engine cycles are captured
only by the Reciprocating Cycles daemons.**

■ *Closed Cycles Ex. #1*: **Simplify the Problem**

An air-standard cycle is. . . Use the systematic approach or the TEST-Map to simplify the problem and reach
the appropriate daemon page ..Closed.Process.Specific.Cycles.

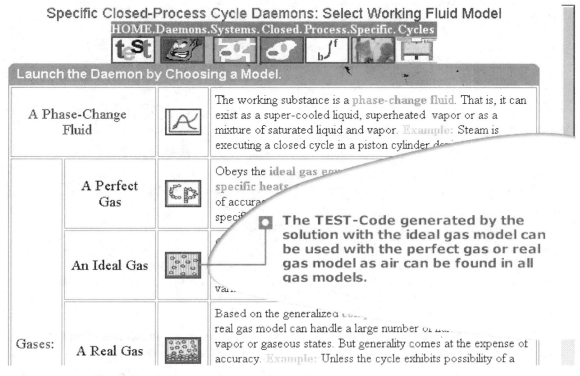

Specific Closed-Process Cycle Daemons: Select Working Fluid Model

HOME.Daemons.Systems. Closed. Process.Specific. Cycles

Launch the Daemon by Choosing a Model.

A Phase-Change Fluid		The working substance is a phase-change fluid. That is, it can exist as a super-cooled liquid, superheated vapor or as a mixture of saturated liquid and vapor. Example: Steam is executing a closed cycle in a piston cylinder de...
	A Perfect Gas	Obeys the ideal gas eq... specific heats... of accura... specifi...
	An Ideal Gas	var...
Gases:	A Real Gas	Based on the generalized con... real gas model can handle a large number of ... vapor or gaseous states. But generality comes at the expense of accuracy. Example: Unless the cycle exhibits possibility of a

The TEST-Code generated by the solution with the ideal gas model can be used with the perfect gas or real gas model as air can be found in all gas models.

■ **Closed Cycles Ex. #1: Idealize – Select a Material Model**

An air-standard cycle is. . . Select the ideal gas model for best accuracy and the daemon is launched.

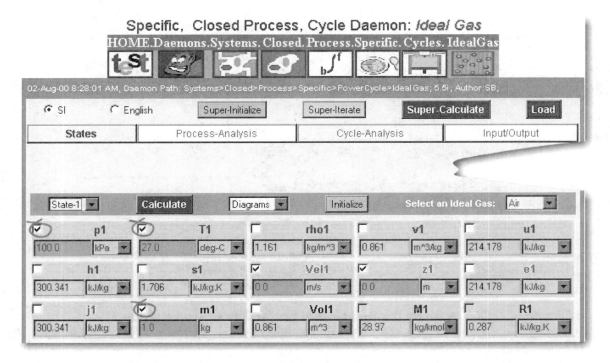

Specific, Closed Process, Cycle Daemon: *Ideal Gas*

HOME.Daemons.Systems. Closed. Process.Specific. Cycles. IdealGas

02-Aug-00 8:28:01 AM, Daemon Path: Systems>Closed>Process>Specific>PowerCycle>IdealGas; 5.5i; Author:SB;

○ SI   ○ English | Super-Initialize | Super-Iterate | **Super-Calculate** | Load |

| **States** | Process-Analysis | Cycle-Analysis | Input/Output |

| State-1 ▼ | **Calculate** | Diagrams ▼ | Initialize | Select an Ideal Gas: | Air ▼ |

☑ p1	☑ T1	☐ rho1	☐ v1	☐ u1
100.0 kPa ▼	27.0 deg-C ▼	1.161 kg/m^3 ▼	0.861 m^3/kg ▼	214.178 kJ/kg ▼
☐ h1	☐ s1	☑ Vel1	☑ z1	☐ e1
300.341 kJ/kg ▼	1.706 kJ/kg.K ▼	0.0 m/s ▼	0.0 m ▼	214.178 kJ/kg ▼
☐ j1	☑ m1	☐ Vol1	☐ M1	☐ R1
300.341 kJ/kg ▼	1.0 kg ▼	0.861 m^3 ▼	28.97 kg/kmol ▼	0.287 kJ/kg.K ▼

■ **IC-Engines Ex. #1: Evaluate the Begin-State of the Compression Stroke (State-1)**

An air-standard cycle is. . . Enter p1, T1, and m1, and Calculate State-1.

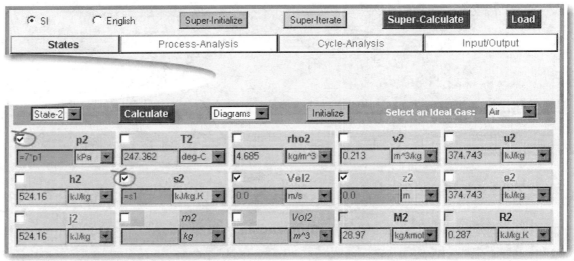

> We can enter m2=m1, or leave it for the mass balance equation to deduce. Similarly, s2=s1, if not entered, could be deduced by the entropy balance equation for the process involving state-1 and state-2.

■ *IC-Engines Ex. #1:* **Evaluate the Finish-State of the Compression Stroke (State-2)**

IC Engines Ex. #1:An air-standard cycle is... Enter p2 and, optionally, s2 and m2. Press the Enter key or the Calculate button.

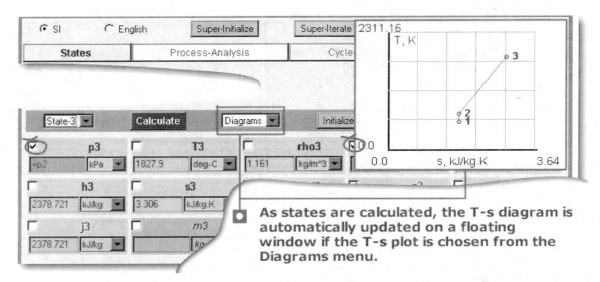

> As states are calculated, the T-s diagram is automatically updated on a floating window if the T-s plot is chosen from the Diagrams menu.

■ *IC-Engines Ex. #1:* **Evaluate the Begin-State for the Power Stroke (State-3)**

An air-standard cycle is... Enter p3 and v3 (= v1), and Calculate. Select p–V or T–s from the pull-down Diagrams menu for appropriate plots.

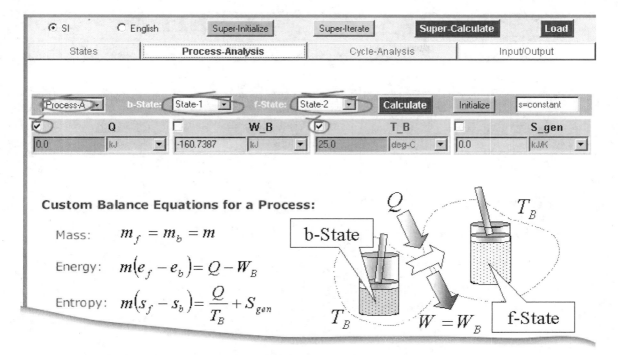

Custom Balance Equations for a Process:

Mass: $m_f = m_b = m$

Energy: $m(e_f - e_b) = Q - W_B$

Entropy: $m(s_f - s_b) = \dfrac{Q}{T_B} + S_{gen}$

- ### *IC-Engines Ex. #1*: Analyze the Compression Stroke (Process-A)

An air-standard cycle is... Select Process-A, load the anchor states, enter Q, and Calculate. The boundary work, W_B, is evaluated.

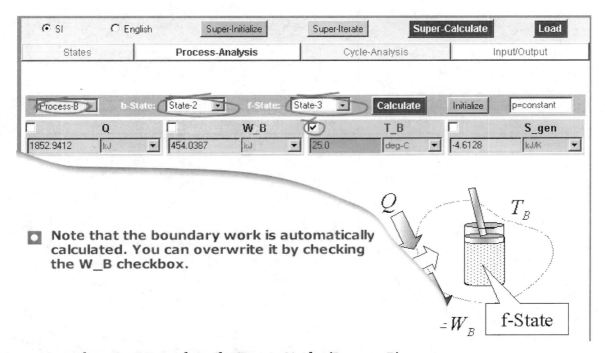

- **Note that the boundary work is automatically calculated. You can overwrite it by checking the W_B checkbox.**

- ### *IC-Engines Ex. #1*: Analyze the Power Stroke (Process-B)

An air-standard cycle is... Select Process-B, load the anchor states, and Calculate. Q and W_B are both evaluated. The impossible value of S_gen indicates that the default value of T_B is unrealistic for heating the gas during the process.

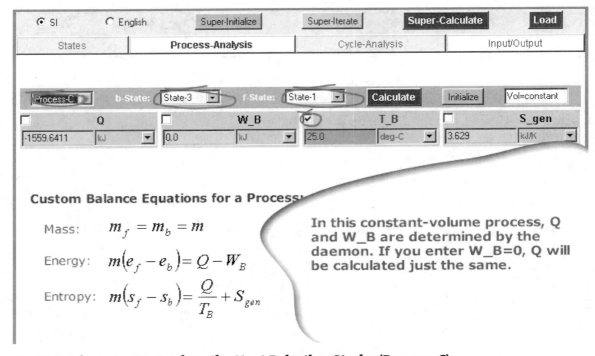

Custom Balance Equations for a Process

Mass: $m_f = m_b = m$

Energy: $m(e_f - e_b) = Q - W_B$

Entropy: $m(s_f - s_b) = \dfrac{Q}{T_B} + S_{gen}$

In this constant-volume process, Q and W_B are determined by the daemon. If you enter W_B=0, Q will be calculated just the same.

■ *IC-Engines Ex. #1:* Analyze the Heat Rejection Stroke (Process-C)

An air-standard cycle is... Select Process-C, load the anchor states, and Calculate. Q and W_B are both evaluated.

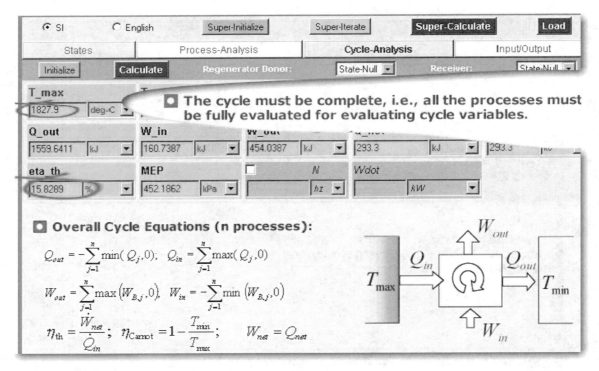

■ Overall Cycle Equations (n processes):

$$Q_{out} = -\sum_{j=1}^{n} \min(Q_j, 0); \quad Q_{in} = \sum_{j=1}^{n} \max(Q_j, 0)$$

$$W_{out} = \sum_{j=1}^{n} \max(W_{B,j}, 0); \quad W_{in} = -\sum_{j=1}^{n} \min(W_{B,j}, 0)$$

$$\eta_{th} = \frac{\dot{W}_{net}}{\dot{Q}_{in}}; \quad \eta_{Carnot} = 1 - \frac{T_{min}}{T_{max}}; \quad W_{net} = Q_{net}$$

■ *IC-Engines Ex. #1:* The Cycle Panel

An air-standard cycle is... Now that all the component processes have been evaluated, the cycle variables are automatically calculated.

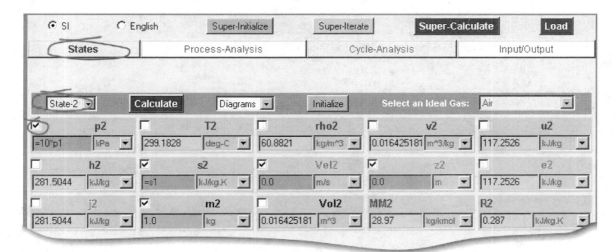

📀 **Note that you must press the Enter key or the Calculate button for any change in parameters to be read by the daemon. The background color of p1 changes from yellow to green.**

■ *IC Engines EX. #1*: **Parametric Study—Change the Pressure Ratio**

An air-standard cycle is... Change the pressure in State-2. Press the Enter key or the Calculate button to register the change, and Super-Calculate.

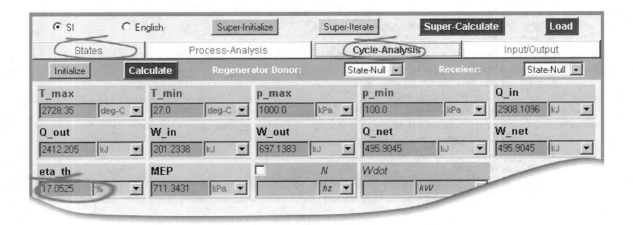

📀 **For a cycle with regeneration, the donor and receiver processes must be selected for appropriate calculation of thermal efficiency.**

■ *IC-Engines Ex. #1*: **Super-Calculate**

An air-standard cycle is... The new efficiency, along with all other cycle variables, such as the mean effective pressure, is evaluated.

```
#####################################################################
# To regenerate this solution, copy the following TEST-Code onto the I/O panel of the
# ..Closed.Process.Specific.Cycles.IdealGas daemon
# and click the Load and Super-Calculate buttons.
#-------------------------------Start of TEST-Code----------------------------------

States {
        State-1: Air;
        Given:      { p1 = 100.0 kPa;  T1 = 27.0 deg-C; Vel1 = 0.0 m/s;
                   z1 = 0.0 m;  m1 = 1.0 kg; }

        State-2: Air;
        Given:      { p2 = "7*p1" kPa;  s2 = "s1" kJ/kg.K; Vel2 = 0.0 m/s; z2 = 0.0 m; }

        State-3: Air;
        Given:      { p3 = "p2" kPa;  v3 = "v1" m^3/kg; Vel3 = 0.0 m/s; z3 = 0.0 m; }
}

Analysis {
        Process-A: b-State = State-1; f-State = State-2;
        Given: { Q = 0.0 kJ; T_B = 25.0 deg-C; }

        Process-B: b-State = State-2; f-State = State-3;
        Given: { T_B = 25.0 deg-C; }

        Process-C: b-State = State-3; f-State = State-1;
        Given: { T_B = 25.0 deg-C; }
}
```

■ *IC-Engines Ex. #1*: **TEST-Code to Regenerate the Visual Solution**

A Diesel cycle has a compression ratio of 18, a cutoff ratio of 2, and a cylinder volume of 2 L. At the beginning of the compression stroke the working fluid is at 95 kPa and 25 deg-C. Determine the thermal efficiency of the cycle by modeling the working fluids as (a) a mixture of ideal gases with the following volumetric composition: N2=70%, O2=10%, and CO2=20%, (b) air with variable specific heat.

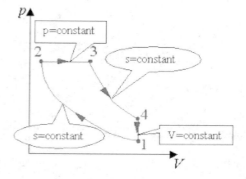

■ **Solution Procedure: Simplify the problem as a specific closed cycle. Use the general mixture model to launch the daemon. Compose the given mixture, evaluate the anchor states, analyze the four strokes, and complete the analysis on the cycle panel.**

■ *Closed Cycles Ex. #1*: **Problem Description**

Specific Closed-Process Cycle Daemons: Select Working Fluid Model

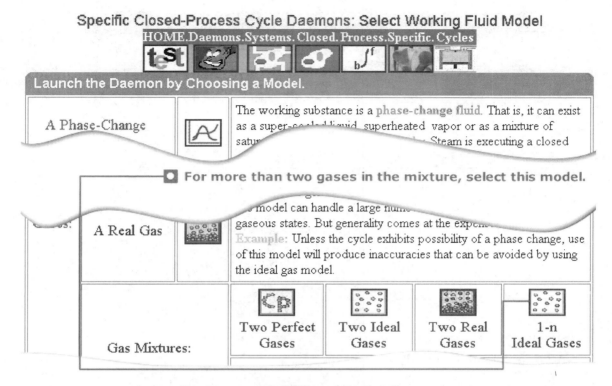

■ *Closed Cycles Ex. #2*: Idealize—Select a Material Model

A Diesel cycle has a compression... Select the 1-n ideal gas mixture model to represent a general mixture.

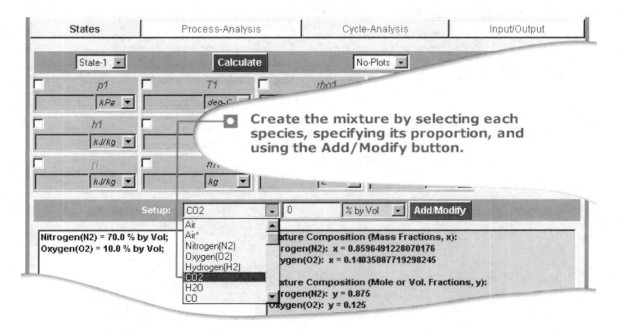

■ *Closed Cycles Ex. #2*: Compose the Mixture

A Diesel cycle has a compression... The mixture needs to be composed only once. To add a species, select it, enter its amount, and click the Add/Modify button. To remove a species, specify its amount to be zero and Add/Modify it.

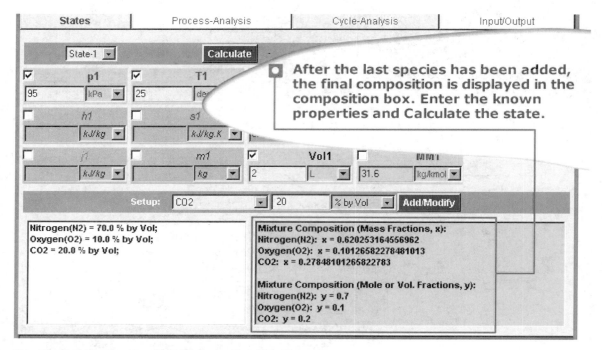

Closed Cycles Ex. #2: Mixture Finalized

A Diesel cycle has a compression. . . The entered amounts are displayed on the left, and the final composition in terms of mass and mole fractions are displayed on the right box.

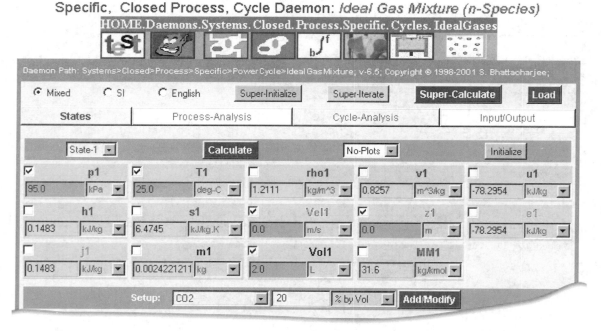

Closed Cycles Ex. #2: Evaluate the b-State of the Compression Stroke (State-1)

A Diesel cycle has a compression. . . Enter p1, T1, and Vol1 and Calculate. Note that the mol mass is a mixture material property.

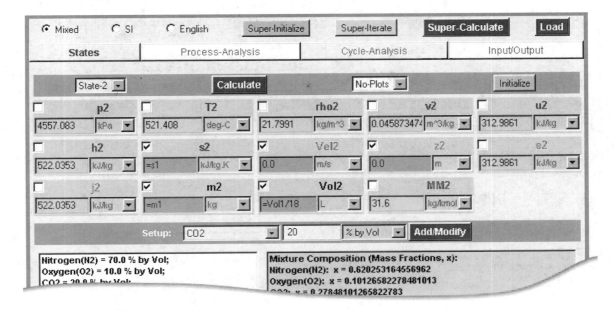

Closed Cycles Ex. #2: Evaluate the f-State of the Compression Stroke (State-2)

A Diesel cycle has a compression... Enter Vol2 (compression ratio is 18), s2 (isentropic), and m2, and Calculate.

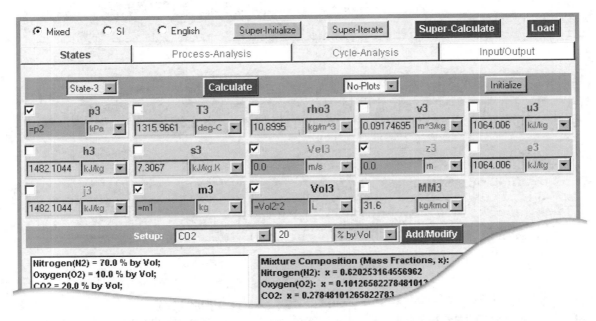

Closed Cycles Ex. #2: Evaluate the f-State of the Combustion Stroke (State-3)

A Diesel cycle has a compression... Enter p3 (constant pressure process), Vol3 (cutoff ratio 2), and m3, and Calculate.

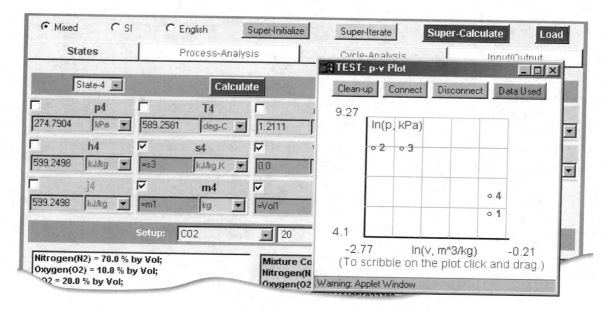

Closed Cycles Ex. #2: Evaluate the f-State of the Power Stroke (State-4)

A Diesel cycle has a compression... Enter Vol4, s4 (isentropic), and m4, and Calculate.

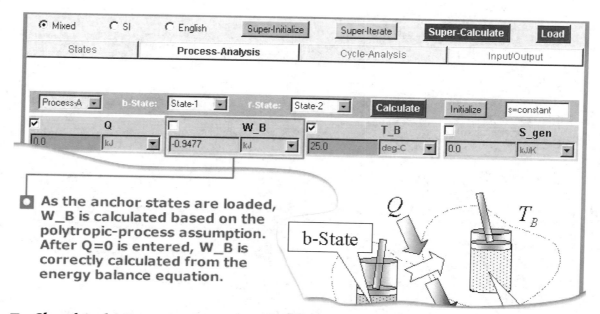

Closed Cycles Ex. #2: Analyze the Compression Stroke (Process-A)

A Diesel cycle has a compression... Select Process-A for the compression stroke. Load the anchor states, State-1 and State-2, enter Q (adiabatic process) and Calculate. The boundary work calculated based on perfect gas assumption is overridden by a more accurate result based on the energy balance equation.

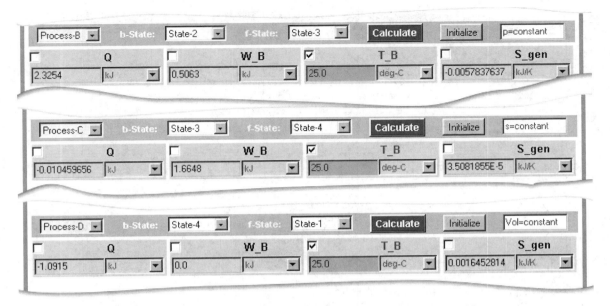

Closed Cycles Ex. #2: Analyze All Other Strokes (Process-B,C, and D)

A Diesel cycle has a compression. . . In a similar way, analyze the combustion, power, and block down strokes.

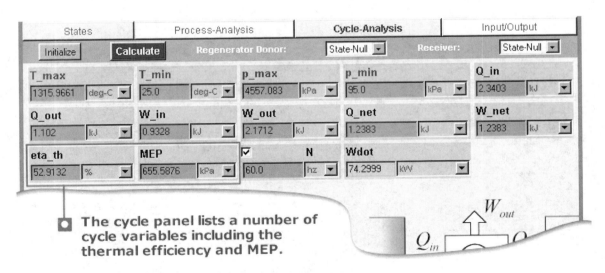

The cycle panel lists a number of cycle variables including the thermal efficiency and MEP.

Closed Cycles Ex. #2: Calculate Cycle Variables

A Diesel cycle has a compression. . . After all the process analysis have been completed, switch to the cycle panel and Calculate all the relevant variables. The thermal efficiency is determined.

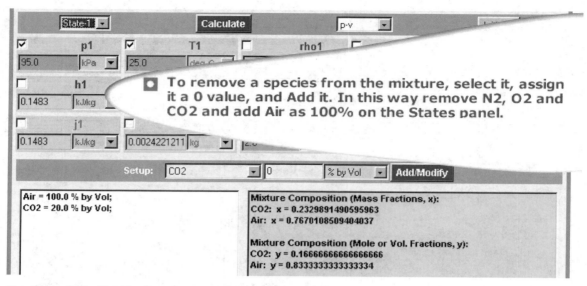

■ *Closed Cycles Ex. #2: Parametric Study*

A Diesel cycle has a compression. . . The effect of a change in parameter can be readily obtained once a visual solution is established. For instance, if the compression ratio is increased to 25 State-2, a Super-Calculate produces eta_th = 57%.

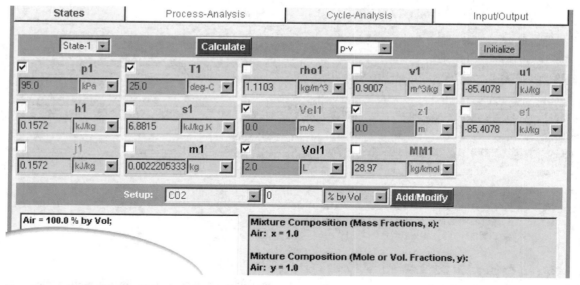

■ *Closed Cycles Ex. #2: Super-Calculate*

A Diesel cycle has a compression. . . Any new mixture can be composed without disturbing the rest of the solution. Once the new mixture is ready, in this case pure air, simply Super-Calculate to update all calculations.

```
#                                              ~ ...xxxxx22523 kJ/K;
#
#      Process-C:  b-State = State-3;  f-State = State-4;
#            Given:  T_B= 25.0 deg-C;
#            Calculated: Q= -0.015160486 kJ;  W_B= 1.783993 kJ;  S_gen= 5.084852E-5 kJ/K;
#
#      Process-D:  b-State = State-4;  f-State = State-1;
#            Given:  T_B= 25.0 deg-C;
#            Calculated: Q= -0.98600614 kJ;  W_B= 0.0 kJ;  S_gen= 0.0014903974 kJ/K;
#
# Cycle:
#---------
#
#            Calculated: T_max= 1503.8727 deg-C;  T_min= 25.0 deg-C;  p_max= 5095.94 kPa;
#                  p_min= 95.0 kPa;  Q_in= 2.3555 kJ;  Q_out= 1.0012 kJ;
#                  W_in= 0.9958 kJ;  W_out= 2.3502 kJ;  Q_net= 1.3544 kJ;
#                  W_net= 1.3544 kJ;  eta_th= 57.4974 %;  MEP= 717.0212 kPa;
#                  N= 60.0 hz;  Wdot= 81.2624 kW;
```

Tab Panel: Buttons on this strip act like tabs.

■ *Closed Cycles Ex. #2:* **New Results**

A Diesel cycle has a compression... All the panels are updated and the results are displayed on the I/O panel. The TEST-Code is also displayed, which can be copied and saved for later use.

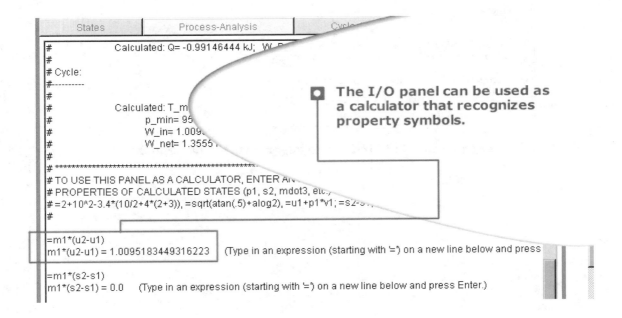

```
States          Process-Analysis          Cycle

#            Calculated: Q= -0.99146444 kJ;  W_B
#
# Cycle:
#---------
#
#            Calculated: T_m
#                  p_min= 95
#                  W_in= 1.009
#                  W_net= 1.3555

#
# ****************************************************
# TO USE THIS PANEL AS A CALCULATOR, ENTER A
# PROPERTIES OF CALCULATED STATES (p1, s2, mdot3, etc.)
# =2+10^2-3.4*(10/2+4*(2+3)), =sqrt(atan(.5)+alog2), =u1+p1*v1; =s2-s
#
```

The I/O panel can be used as a calculator that recognizes property symbols.

```
=m1*(u2-u1)
m1*(u2-u1) = 1.0095183449316223    (Type in an expression (starting with '=') on a new line below and press

=m1*(s2-s1)
m1*(s2-s1) = 0.0    (Type in an expression (starting with '=') on a new line below and press Enter.)
```

■ *Closed Cycles Ex. #2:* **i/O Panel as a Calculator**

A Diesel cycle has a compression... As with all other daemons, the I/O panel can be used as a scientific calculator.

```
####################################################################
# To regenerate this solution, copy the following TEST-Code onto the I/O panel of the
# ..Closed.Process.Specific.Cycles.n-IdealGases daemon,
# click Load, manually create the mixture, and Super-Calculate.
#--------------------Start of TEST-Code----------------------------------------------------------------
----
 Mixture {
   Nitrogen(N2) = 70.0 % by Vol;  Oxygen(O2) = 10.0 % by Vol;  CO2 = 20.0 % by Vol;
 }

 States {
 State-1:  mixture;
 Given:        { p1= 95.0 kPa;  T1= 25.0 deg-C;  Vel1= 0.0 m/s;  z1= 0.0 m;  Vol1 = 2.0 L; }

 State-2:  mixture;
 Given:    { s2= "s1" kJ/kg.K;  Vel2= 0.0 m/s;  z2= 0.0 m;  m2= "m1" kg;  Vol2=
 "Vol1/18" L; }

 State-3:  mixture;
 Given:        { p3= "p2" kPa;  Vel3= 0.0 m/s;  z3= 0.0 m;  m3= "m1" kg;  Vol3= "Vol2*2"
 L; }

 State-4:  mixture;
 Given:        { s4= "s3" kJ/kg.K;  Vel4= 0.0 m/s;  z4= 0.0 m;  m4= "m1" kg;  Vol4= "Vol1"
 L; }
 }

 Analysis {
 Process-A: b-State = State-1;  f-State = State-2;
 Given: { Q= 0.0 kJ;  T_B= 25.0 deg-C; }

 Process-B: b-State = State-2;  f-State = State-3;
 Given: { T_B= 25.0 deg-C; }

 Process-C: b-State = State-3;  f-State = State-4;
 Given: { Q= 0.0 kJ;  T_B= 25.0 deg-C; }

 Process-D: b-State = State-4;  f-State = State-1;
 Given: { T_B= 25.0 deg-C; }
 }
```

■ *IC-Engines Ex. #1*: **TEST-Code to Regenerate Visual Solution**

Open-Device Cycles

Power and refrigeration cycles, such as the Rankine cycle, Brayton cycle, or their reverse counterparts, are made by connecting open devices in a closed network. To emphasize that fact, the daemons are called Open-Device Cycle daemons or simply **Open Cycle** daemons. They are constructed by adding a Cycle panel on top of the Multi-flow, Open, Steady Device daemon because of its versatility. Such a device can be easily manipulated to act like a single-flow device or a multi-flow device with or without mixing. After all the devices of a cycle are analyzed, the **Cycle Panel** puts them together and extracts all the cycle-related variables, after checking to make sure that the cycle is complete.

The Device daemons covered in Chapter 4 are the keys to master the Open Cycle daemons. Although only a few examples are given here, the Tutorial and Chapter 9 and 10 in the Archive contain a number of examples with different types of reciprocating cycles.

Example: In a steam power plant operating on the ideal regenerative Rankine cycle with one open feedwater heater, steam enters the turbine at 15 MPa and 620 deg C and is condensed in the condenser at a pressure of 15 kPa. Bleeding from the turbine to the FWH occurs at 1 MPa.

Determine (a) the fraction of steam extracted and (b) the thermal efficiency of the cycle. (c) How would the efficiency change if the bleeding were stopped altogether.

■ Solution Procedure: Simplify the problem as an open, specific, power cycle problem. Evaluate all seven states as best as possible. Analyze each device and then Super-Calculate the complete solution. Change any input parameter and Super-Calculate to update the entire solution.

■ *Steam Power Ex. #1*: **Problem Description**

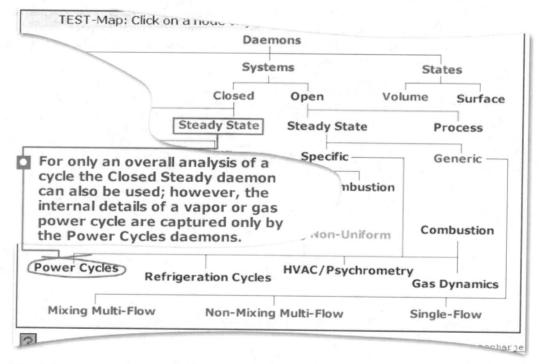

■ *Steam Power Ex. #1*: **Simplify the Problem**

In a steam power plant operating... Use the systematic approach or this TEST-Map to get to the right daemon page:... Open.Steady.Specific.Power Cycles.

Open Steady Power Cycle Daemons: Select Working Fluid Model
HOME.Daemons.Systems. Open. SteadyState.Specific.PowerCycles

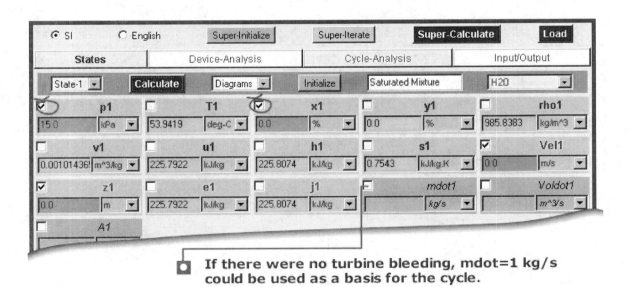

Launch the Daemon by Choosing a Model.

A Phase-Change Fluid		The working substance is a phase-change fluid. That is, it can exist as a super-cooled liquid, superheated vapor or as a mixture of saturated liquid and vapor. Example: H2O executing the Rankine cycle.
	A Perfect Gas	Obeys the ideal gas equation (pv=RT). Moreover the specific heats remain constant - a simplification at the expense of accuracy. Example: The air standard Brayton cycle.
	An Ideal Gas	Obeys the ideal gas equation (pv=RT). Specific heats are temperature dependent; thus, the model is more accurate than the perfect gas model. Example: The air standard Brayton cycle with variable specific heats for better accuracy.
		Based on the generalized compressibility chart (pv=ZRT)

■ *Steam Power Ex. #1*: **Select a Material Model**

In a steam power plant operating. . . The phase-change model is the obvious choice for modeling steam.

If there were no turbine bleeding, mdot=1 kg/s could be used as a basis for the cycle.

■ *Steam Power Ex. #1*: **Evaluate the Feedwater Pump Inlet State (State-1)**

In a steam power plant operating. . . Select State-1, enter p1 and x1 (x1 = 0 makes the liquid saturated). Note that `mdot` is an unknown at this point.

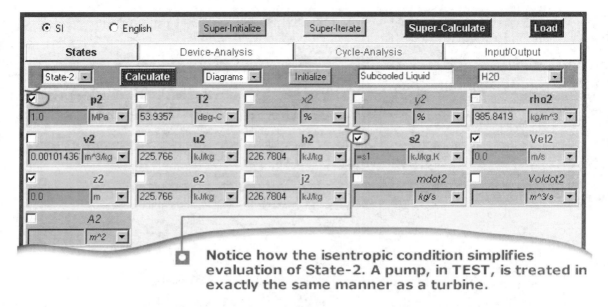

Notice how the isentropic condition simplifies evaluation of State-2. A pump, in TEST, is treated in exactly the same manner as a turbine.

■ *Steam Power Ex. #1*: Evaluate the Feedwater Pump Exit State (State-2)

In a steam power plant operating... Select State-2 as the first pump exit. Enter p2 and s2 (= s) for the isentropic pump. Calculate.

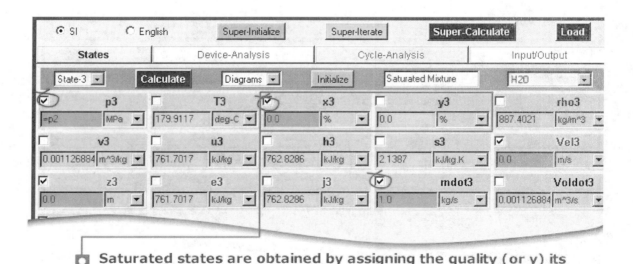

Saturated states are obtained by assigning the quality (or y) its maximum (100% for saturated vapor) or minimum (0% for saturated liquid) value.

■ *Steam Power Ex. #1*: Evaluate the Feedwater Heater Exit State (State-3)

In a steam power plant operating... Select State-3 as the feedwater heater exit. Enter p3 and x3. Note that mdot3, the total mass flow rate, is assumed to be 1 kg/s.

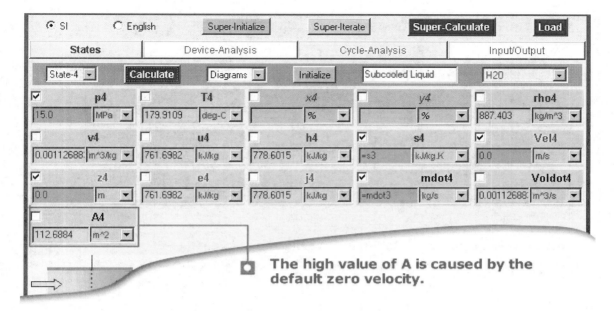

The high value of A is caused by the default zero velocity.

Steam Power Ex. #1: Evaluate the Boiler Inlet State (State-4)

In a steam power plant operating... Select State-4 as the second pump exit. Enter p4, s4 and mdot4. Calculate.

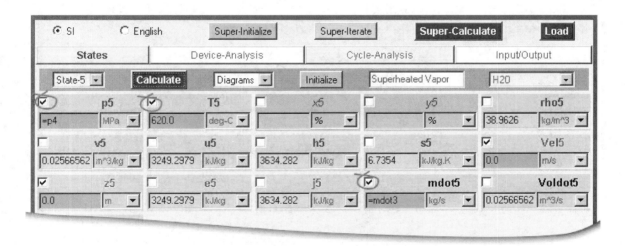

Steam Power Ex. #1: Evaluate the Boiler Exit State (State-5)

Steam Power Ex. #1: In a steam power plant operating... Select State-5 as the boiler exit. Enter p5, T5, and mdot5. Calculate.

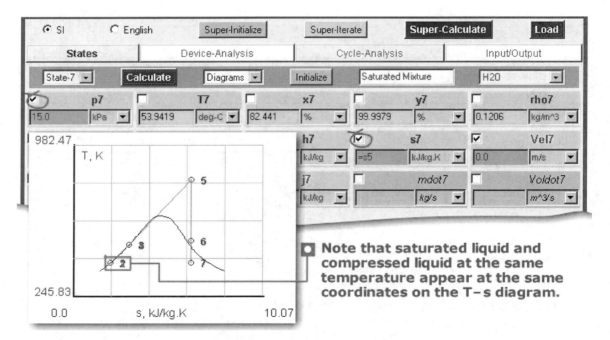

| | SI | | English | | Super-Initialize | | Super-Iterate | | Super-Calculate | | Load |

| States | | Device-Analysis | | Cycle-Analysis | | Input/Output |

State-6 ▾ | Calculate | Diagrams ▾ | Initialize | Superheated Vapor | H2O ▾

| ☑ | p6 | □ | T6 | □ | x6 | □ | y6 | □ | rho6 |
| =p2 | MPa ▾ | 209.0023 | deg-C ▾ | | % ▾ | | % ▾ | 4.7446 | kg/m^3 ▾ |

| □ | v6 | □ | u6 | □ | h6 | ☑ | s6 | ☑ | Vel6 |
| 0.2108 | m^3/kg ▾ | 2637.7285 | kJ/kg ▾ | 2848.494 | kJ/kg ▾ | =s5 | kJ/kg.K ▾ | 0.0 | m/s ▾ |

| ☑ | z6 | □ | e6 | □ | j6 | □ | mdot6 | □ | Voldot6 |
| 0.0 | m ▾ | 2637.7285 | kJ/kg ▾ | 2848.494 | kJ/kg ▾ | | kg/s ▾ | 0.043101545 | m^3/s ▾ |

■ *Steam Power Ex. #1*: **Evaluate the Turbine Bleeding State (State-6)**

Steam Power Ex. #1: In a steam power plant operating. . . Select State-6 as the first turbine exit. Enter p6 and s6. Calculate.

| | SI | | English | | Super-Initialize | | Super-Iterate | | Super-Calculate | | Load |

| States | | Device-Analysis | | Cycle-Analysis | | Input/Output |

State-7 ▾ | Calculate | Diagrams ▾ | Initialize | Saturated Mixture | H2O ▾

| ☑ | p7 | □ | T7 | □ | x7 | □ | y7 | □ | rho7 |
| 15.0 | kPa ▾ | 53.9419 | deg-C ▾ | 82.441 | % ▾ | 99.9979 | % ▾ | 0.1206 | kg/m^3 ▾ |

| | | | | | h7 | ☑ | s7 | ☑ | Vel7 |
| | | | | | kJ/kg ▾ | =s5 | kJ/kg.K ▾ | 0.0 | m/s ▾ |

| | | | | | j7 | □ | mdot7 | □ | Voldot7 |
| | | | | | kJ/kg ▾ | | kg/s ▾ | | m^3/s ▾ |

○ **Note that saturated liquid and compressed liquid at the same temperature appear at the same coordinates on the T–s diagram.**

■ *Steam Power Ex. #1*: **Evaluate the Turbine Exit State (State-7)**

Steam Power Ex. #1: In a steam power plant operating. . . Select State-7 as the second turbine exit. Enter p7 and s7. Calculate. Plot the T–s diagram to check for consistency in state evaluation.

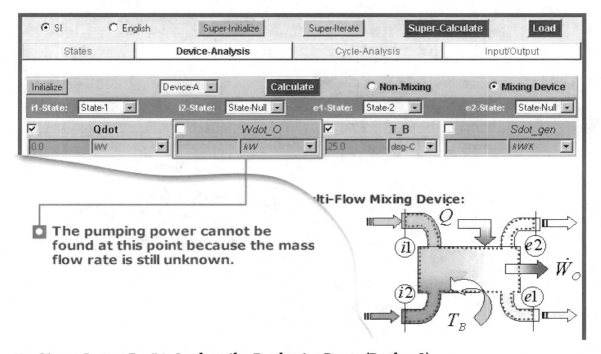

Steam Power Ex. #1: Analyze the Feedwater Pump (Device-A)

In a steam power plant operating... Select Device-A as the first pump. Load the inlet and exit states, enter Qdot, and Calculate. The pumping power cannot be determined until the heater is analyzed and a Super-Calculate updates all calculations.

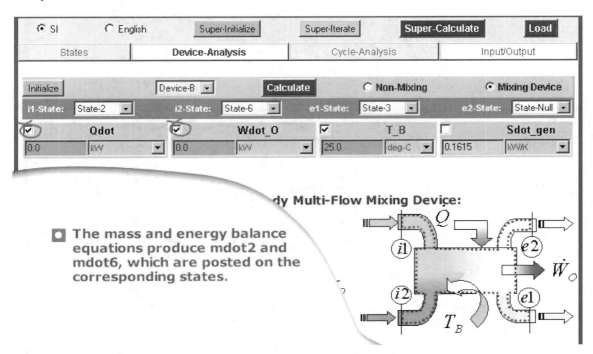

Steam Power Ex. #1: Analyze the Feedwater Heater (Device-B)

In a steam power plant operating... Select Device-B as the feedwater heater. Load the inlet and exit states, enter Qdot and Wdot_O, check the Mixing Device radio-button, and Calculate. The bleeding rate of steam is posted back to States 2 and 3.

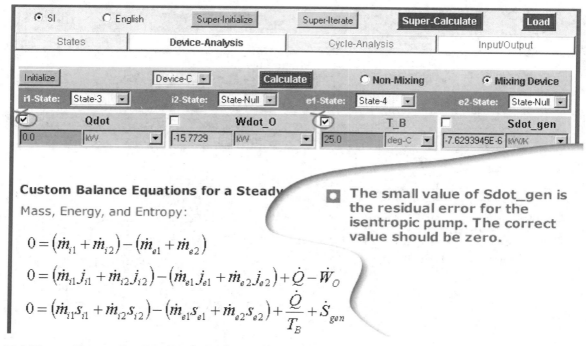

Custom Balance Equations for a Steady

Mass, Energy, and Entropy:

$$0 = (\dot{m}_{i1} + \dot{m}_{i2}) - (\dot{m}_{e1} + \dot{m}_{e2})$$

$$0 = (\dot{m}_{i1} j_{i1} + \dot{m}_{i2} j_{i2}) - (\dot{m}_{e1} j_{e1} + \dot{m}_{e2} j_{e2}) + \dot{Q} - \dot{W}_O$$

$$0 = (\dot{m}_{i1} s_{i1} + \dot{m}_{i2} s_{i2}) - (\dot{m}_{e1} s_{e1} + \dot{m}_{e2} s_{e2}) + \frac{\dot{Q}}{T_B} + \dot{S}_{gen}$$

The small value of Sdot_gen is the residual error for the isentropic pump. The correct value should be zero.

■ *Steam Power Ex. #1*: Analyze the Boiler Pump

In a steam power plant operating... Select Device-C as the second pump. Load the inlet and exit states, enter Qdot, and Calculate. Now that the mass flow rates are known, the pumping power is determined.

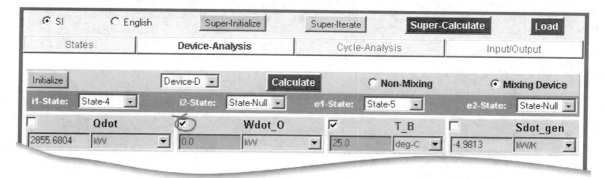

The negative value of Sdot_gen results from the unrealistic default value of T_B for the steam generator. Heat cannot be supplied to the steam generator if the surroundings are at a temperature lower than the steam.

■ *Steam Power Ex. #1*: Analyze the Boiler (Device-D)

Steam Power Ex. #1: In a steam power plant operating... Select Device-D as the steam generator. Load the inlet and exit states, enter Wdot_O, and Calculate. The heat transfer is determined.

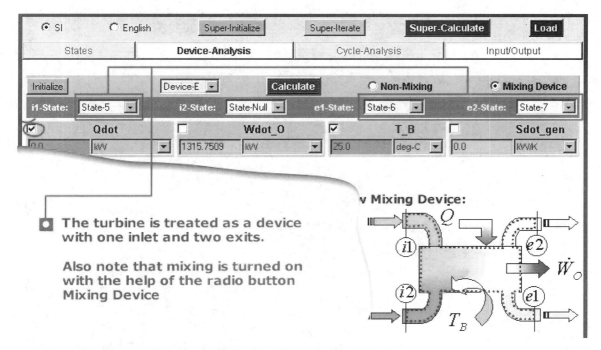

The turbine is treated as a device with one inlet and two exits.

Also note that mixing is turned on with the help of the radio button Mixing Device

■ *Steam Power Ex. #1:* **Analyze the Turbine (Device-E)**

In a steam power plant operating ... Select Device-E as the steam turbine. Load the inlet and exit states, enter Qdot, and Calculate. Now that the mass flow rates are known, the turbine output is determined.

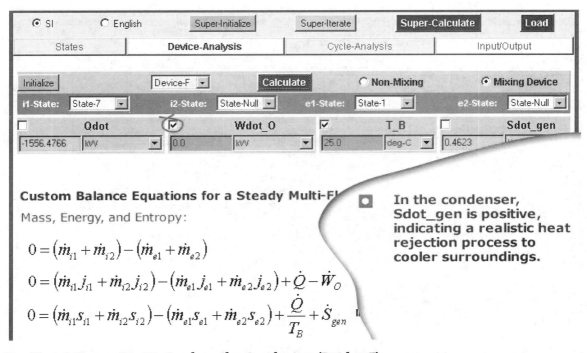

Custom Balance Equations for a Steady Multi-Fl

Mass, Energy, and Entropy:

$$0 = \left(\dot{m}_{i1} + \dot{m}_{i2} \right) - \left(\dot{m}_{e1} + \dot{m}_{e2} \right)$$

$$0 = \left(\dot{m}_{i1} j_{i1} + \dot{m}_{i2} j_{i2} \right) - \left(\dot{m}_{e1} j_{e1} + \dot{m}_{e2} j_{e2} \right) + \dot{Q} - \dot{W}_O$$

$$0 = \left(\dot{m}_{i1} s_{i1} + \dot{m}_{i2} s_{i2} \right) - \left(\dot{m}_{e1} s_{e1} + \dot{m}_{e2} s_{e2} \right) + \frac{\dot{Q}}{T_B} + \dot{S}_{gen}$$

In the condenser, Sdot_gen is positive, indicating a realistic heat rejection process to cooler surroundings.

■ *Steam Power Ex. #1:* **Analyze the Condenser (Device-F)**

In a steam power plant operating ... Select Device-F as the condenser. Load the inlet and exit states, enter Wdot_O, and Calculate. The heat transfer is determined.

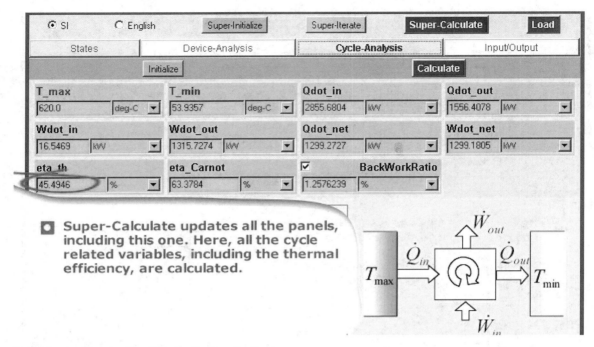

> **Super-Calculate updates all the panels, including this one. Here, all the cycle related variables, including the thermal efficiency, are calculated.**

Steam Power Ex. #1: Cycle Analysis

In a steam power plant operating. . . Super-Calculate iterates between the States and Analysis panels. The thermal efficiency is found. If more iterations are necessary, use the Super-Iterate button.

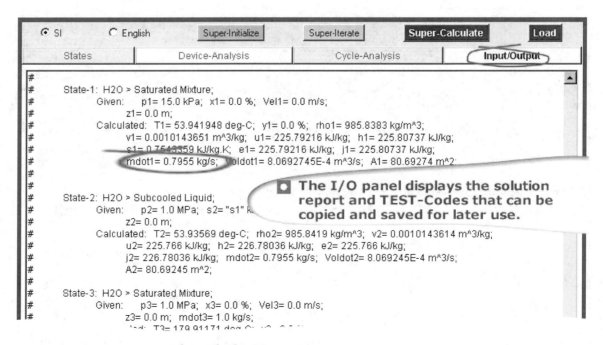

> **The I/O panel displays the solution report and TEST-Codes that can be copied and saved for later use.**

Steam Power Ex. #1: The Solution Report

In a steam power plant operating. . . Super-Calculate also produces a detailed report on the I/O panel.

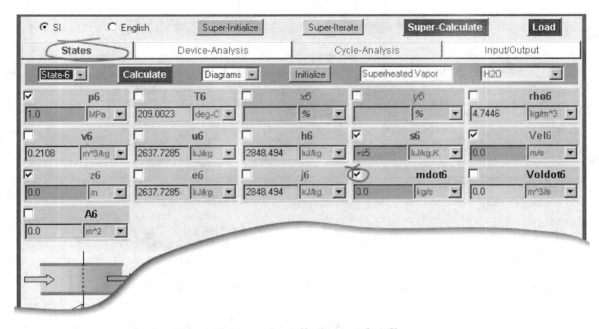

Steam Power Ex. #1: What-If Scenario—Eliminate Bleeding

In a steam power plant operating... The bleeding flow rate is reduced to zero by assigning mdot6 = 0. Alternatively, we could have modified mdot7 to 1 kg/s in State-7.

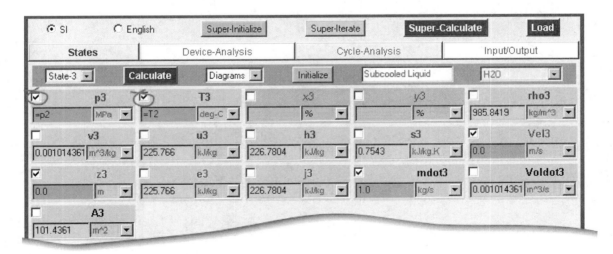

Steam Power Ex. #1: What-If Scenario—Eliminate Bleeding

In a steam power plant operating... In the absence of bleeding, State-3 must be identical to State-2, which is achieved here by equating p3 and T3 to p2 and T2, respectively.

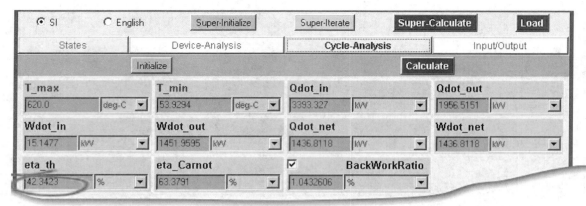

The thermal efficiency is reduced by about 3% without the feedwater heater. To perform a parametric study, say, on the bleed pressure, copy the TEST-Code into the I/O panel, load and Super-Calculate to reproduce this solution. Next, change p2 and update solution using Super-Calculate.

Steam Power Ex. #1: Super-Calculate to Find the New Efficiency

In a steam power plant operating... Update all calculations using Super-Calculate. In the absence of bleeding, the efficiency decreases by about 3%.

```
####################################################################
# To regenerate this solution, copy the following TEST-Code onto the I/O panel of the
# ..Open.Steady.Specific.PowerCycle.PhaseChanger daemon
# and click the Load and Super-Calculate buttons.
#-------------------Start of TEST-Code--------------------------------------------------------
----

   States {
   State-1:  H2O;
   Given:        { p1= 15.0 kPa;  x1= 0.0 %;  Vel1= 0.0 m/s;  z1= 0.0 m;  }

   State-2:  H2O;
   Given:        { p2= 1.0 MPa;  s2= "s1" kJ/kg.K;  Vel2= 0.0 m/s;  z2= 0.0 m;  }

   State-3:  H2O;
   Given:        { p3= "p2" MPa;  x3= 0.0 %;  Vel3= 0.0 m/s;  z3= 0.0 m;  mdot3= 1.0 kg/s;  }

   State-4:  H2O;
   Given:        { p4= 15.0 MPa;  s4= "s3" kJ/kg.K;  Vel4= 0.0 m/s;  z4= 0.0 m;  mdot4=
   "mdot3" kg/s;  }

   State-5:  H2O;
   Given:        { p5= "p4" MPa;  T5= 620.0 deg-C;  Vel5= 0.0 m/s;  z5= 0.0 m;  mdot5=
   "mdot3" kg/s;  }

   State-6:  H2O;
   Given:        { p6= "p2" MPa;  s6= "s5" kJ/kg.K;  Vel6= 0.0 m/s;  z6= 0.0 m;  }

   State-7:  H2O;
   Given:        { p7= 15.0 kPa;  s7= "s5" kJ/kg.K;  Vel7= 0.0 m/s;  z7= 0.0 m;  }
   }

   Analysis {
   Device-A: i-State = State-1;  e-State = State-2;  Mixing: true
   Given: { Qdot= 0.0 kW;  T_B= 25.0 deg-C;  }

   Device-B: i-State = State-2, State-6;  e-State = State-3;  Mixing: true
   Given: { Qdot= 0.0 kW ;  Wdot_O= 0.0 kW;  T_B= 25.0 deg-C;  }

   Device-C: i-State = State-3;  e-State = State-4;  Mixing: true
   Given: { Qdot= 0.0 kW;  T_B= 25.0 deg-C;  }

   Device-D: i-State = State-4;  e-State = State-5;  Mixing: true
   Given: { Wdot_O= 0.0 kW;  T_B= 25.0 deg-C;  }

   Device-E: i-State = State-5;  e-State = State-6, State-7;  Mixing: true
   Given: { Qdot= 0.0 kW;  T_B= 25.0 deg-C;  }

   Device-F: i-State = State-7;  e-State = State-1;  Mixing: true
   Given: { Wdot_O= 0.0 kW;  T_B= 25.0 deg-C;  }

   }
```

■ *Steam Power Ex. #1*: **TEST-Code to Regenerate the Visual Solution**

Example: A Brayton cycle with regeneration and air as the working fluid operates on a pressure ratio of 8. The minimum and maximum temperatures of the cycle are 300 and 1200 K. The adiabatic efficiencies of the turbine and the compressor are 80% and 82%, respectively. The regenerator effectiveness is 65%. Determine (a) the thermal efficiency and (b) net power output. Assume variable c_p.

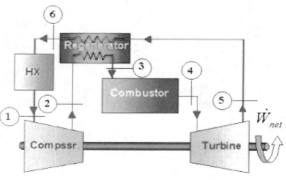

How would the answers change if the regenerator efficiency were increased to 100%?

> ⊙**Solution Procedure: Simplify the problem as an open, specific, power cycle problem with ideal gas as the working fluid. In evaluating the states, use algebraic relations as much as possible. Analyze each device individually. Evaluate the unknowns with a Super-Calculate. Change any parameter, and Super-Calculate to update the solution.**

■ **Gas Turbine Ex. #1: Problem Description**

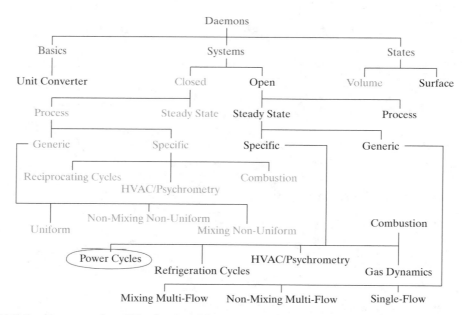

■ **Gas Turbine Ex. #1: Simplify the Problem**

A Brayton cycle with regeneration. . . The appropriate simplification for this problem leads to the following daemon page: . . . Open.Steady.Specific.Power Cycles.

Open Steady Power Cycle Daemons: Select Working Fluid Model

HOME.Daemons.Systems. Open. SteadyState.Specific.PowerCycles

Launch the Daemon by Choosing a Model.

A Phase-Change Fluid		The working substance is a phase-change fluid. That is, it can exist as a super-cooled liquid, superheated vapor or as a mixture of saturated liquid and vapor. Example: H2O executing the Rankine cycle.
	A Perfect Gas	Obeys the ideal gas equation (pv=RT). Moreover the specific heats remain constant - a simplification at the expense of accuracy. Example: The air standard Brayton cycle.
	An Ideal Gas	Obeys the ideal gas equation (pv=RT). Specific heats are temperature dependent; thus, the model is more accurate than the perfect gas model. Example: The air standard Brayton cycle with variable specific heats for better accuracy.
		Based on the generalized compressibility chart (pv=ZRT

■ ***Gas Turbine Ex. #1*: Simplify the Problem**

A Brayton cycle with regeneration... Choose Ideal Gas to model air with variable specific heat.

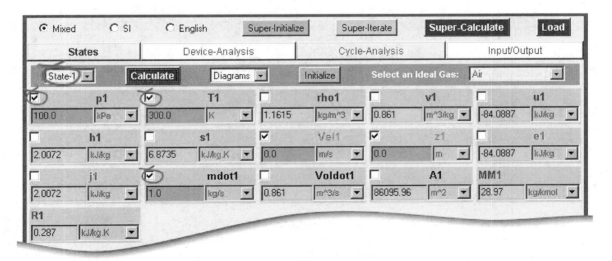

■ ***Gas Turbine Ex. #1*: Evaluate the Compressor Inlet State (State-1)**

A Brayton cycle with regeneration... Base the solution on a mass flow rate of 1 kg/s. Enter p1 and T1, and Calculate the compressor inlet state.

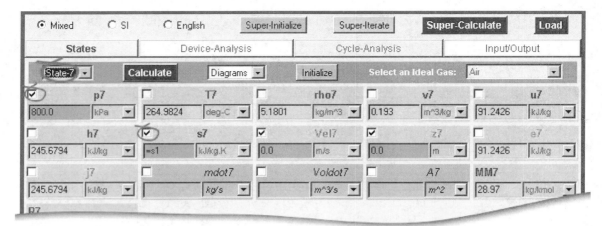

■ State-7 is isentropic to State-1.

■ *Gas Turbine Ex. #1*: **Evaluate the Isentropic Compressor Exit State (State-7)**

A Brayton cycle with regeneration. . . Select an unused state number (State-7) to represent the isentropic state. Enter p7 and s7. Calculate.

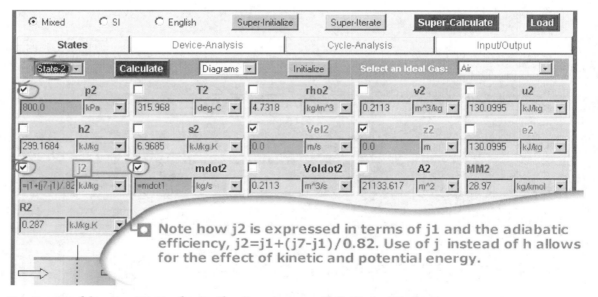

■ Note how j2 is expressed in terms of j1 and the adiabatic efficiency, j2=j1+(j7-j1)/0.82. Use of j instead of h allows for the effect of kinetic and potential energy.

■ *Gas Turbine Ex. #1*: **Evaluate the Compressor Exit State (State-2)**

A Brayton cycle with regeneration. . . Use adiabatic efficiency to relate j2 with j1 and j7 (use of j instead of h allows us to use non-zero velocity, if necessary). Enter p2 and mdot2, and Calculate.

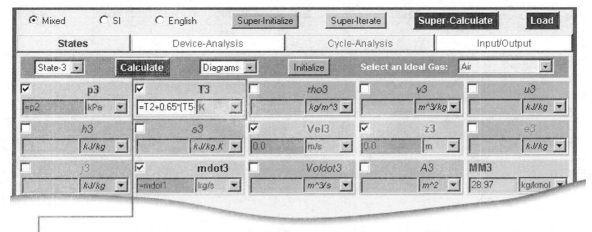

○ The state is not completely evaluated, because T3, expressed in terms of State-5 as T2+0.65*(T5-T2), is not yet known. After all the known information are entered, we can use Super-Calculate to complete all unfinished state calculations.

■ *Gas Turbine Ex. #1*: Evaluate the Combustor Inlet State (State-3)

A Brayton cycle with regeneration... Use regenerator effectiveness to relate T3 with T2 and T5. The white background of T3 indicates that the expression cannot be evaluated, due to format error or unknown variables (T5 in this case).

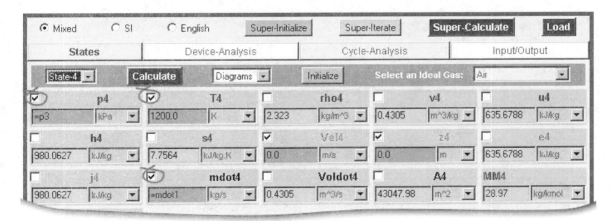

○ State-4 is the turbine inlet state and is straightforward to determine.

■ *Gas Turbine Ex. #1*: Evaluate the Turbine Inlet State (State-4)

A Brayton cycle with regeneration... Enter turbine inlet pressure p4 (same as compressor exit pressure with no loss in the regenerator or combustor), T4, and mdot4, and Calculate (or press the Enter key).

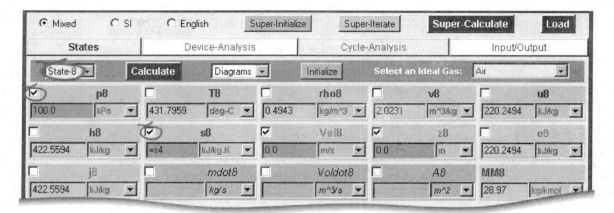

○ State-8 is the isentropic turbine exit state.

■ *Gas Turbine Ex. #1*: **Evaluate the Isentropic Turbine Exit State (State-8)**

A Brayton cycle with regeneration. . . Pick an unused state number, State-8. Enter p8 and s8, isentropic to State-4. Calculate.

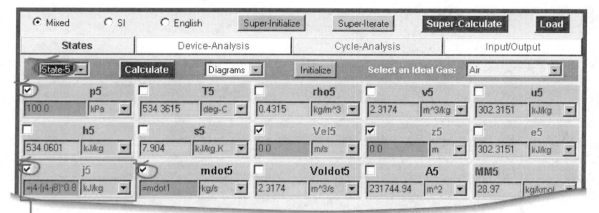

○ An algebraic expression is preferable to an absolute value to facilitate parametric study.

■ *Gas Turbine Ex. #1*: **Evaluate the Turbine Exit State (State-5)**

A Brayton cycle with regeneration. . . Relate j5 with j8 and j4 through the adiabatic turbine efficiency. Enter p5 and mdot5. Calculate.

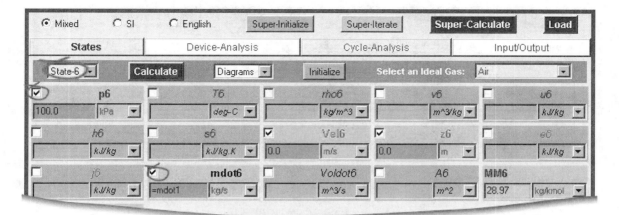

O **State-6 cannot be evaluated without analyzing the regenerator.**

■ *Gas Turbine Ex. #1*: **Evaluate the Regenerator Exit State (State-6)**

A Brayton cycle with regeneration... Enter p6 and mdot6. Calculate. The state cannot be evaluated until a device analysis is performed on the regenerator.

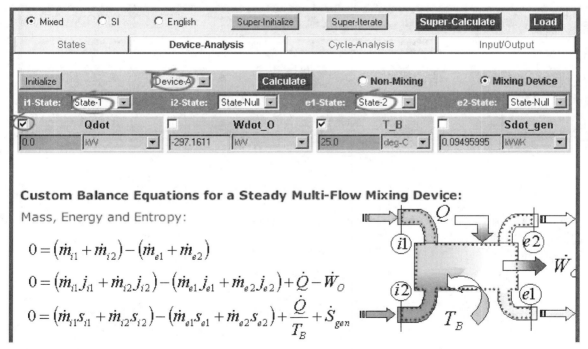

Custom Balance Equations for a Steady Multi-Flow Mixing Device:

Mass, Energy and Entropy:

$$0 = \left(\dot{m}_{i1} + \dot{m}_{i2}\right) - \left(\dot{m}_{e1} + \dot{m}_{e2}\right)$$

$$0 = \left(\dot{m}_{i1} j_{i1} + \dot{m}_{i2} j_{i2}\right) - \left(\dot{m}_{e1} j_{e1} + \dot{m}_{e2} j_{e2}\right) + \dot{Q} - \dot{W}_O$$

$$0 = \left(\dot{m}_{i1} s_{i1} + \dot{m}_{i2} s_{i2}\right) - \left(\dot{m}_{e1} s_{e1} + \dot{m}_{e2} s_{e2}\right) + \frac{\dot{Q}}{T_B} + \dot{S}_{gen}$$

■ *Gas Turbine Ex. #1*: **Analyze the Compressor (Device-A)**

A Brayton cycle with regeneration... Select Device-A to represent the compressor. Load the i- and e-states. Enter Qdot. Calculate to find the compressor power.

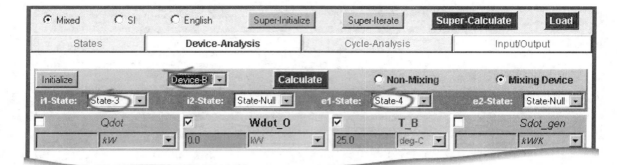

◻ **Qdot in the combustor cannot be determined, because State-3 is not yet evaluated.**

■ *Gas Turbine Ex. #1:* **Analyze the Combustor (Device-B)**

A Brayton cycle with regeneration. . . Select Device-B to represent the combustor. Load the i- and e-states. Enter Wdot_O. Calculate. We have to wait for State-3 to be completely evaluated before Qdot can be determined.

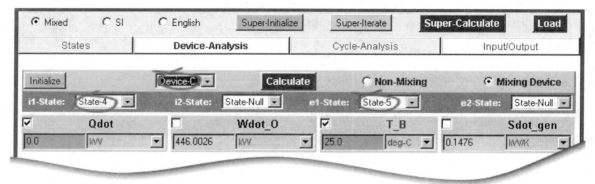

◻ **Device-C is the turbine.**

■ *Gas Turbine Ex. #1:* **Analyze the Turbine (Device-C)**

A Brayton cycle with regeneration. . . Select Device-C to represent the turbine. Load the i- and e-states. Enter Qdot and Calculate. The power produced by the turbine is determined.

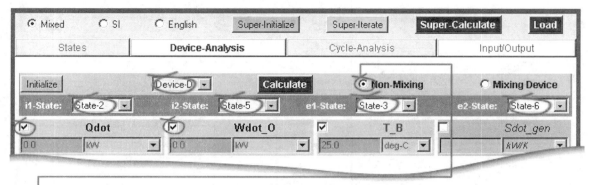

■ Device-D is the regenerator. Notice the selection of the Non-Mixing category. Solutions are posted in State-3 and State-6.

■ *Gas Turbine Ex. #1*: Analyze the Regeneration (Device-D)

A Brayton cycle with regeneration... Select Device-D to represent the regenerator. Choose the Non-Mixing option. You will see the device image and balance equations adjust to this option. Load the i- and e-states, and enter Qdot and Wdot_O. Calculate. The solution of the energy balance equation produces j6, which is posted back to State-6.

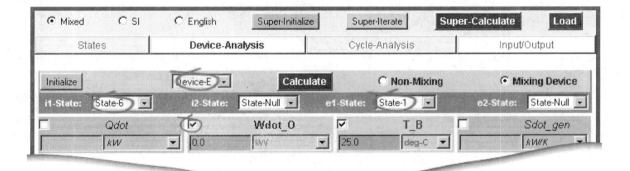

■ Device-E is the heat exchanger (to complete the air-standard cycle). Once again, we need Super-Calculate to evaluate Qdot.

■ *Gas Turbine Ex. #1*: Analyze the Heat Exchanger (Device-E)

A Brayton cycle with regeneration... Select Device-E to represent the heat exchanger. Load the i- and e-states. Enter Wdot_O and Calculate. Once again, Super-Calculate is necessary to evaluate Qdot.

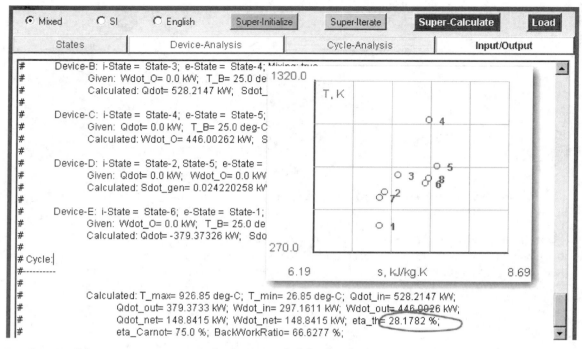

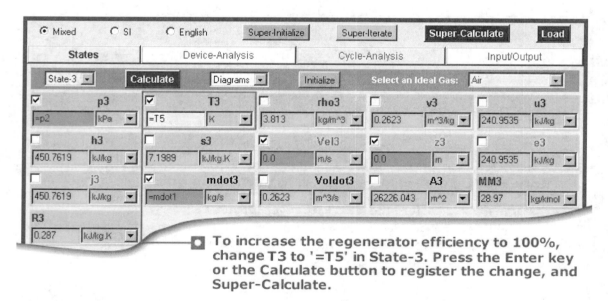

■ *Gas Turbine Ex. #1*: Super-Calculate to Finish the Solution

A Brayton cycle with regeneration... Super-Calculate produces a detailed report that includes the efficiency and the net power. Sometimes Super-Iterate may be necessary in a complex problem for sufficient iterations between the States and Analysis panel. The T–s diagram shown here is produced in the States panel.

> **☐** To increase the regenerator efficiency to 100%, change T3 to '=T5' in State-3. Press the Enter key or the Calculate button to register the change, and Super-Calculate.

■ *Gas Turbine Ex. #1*: Parametric Study—Change Regenerator Effectiveness

A Brayton cycle with regeneration... To change the regenerator effectiveness, go back to State-3 and change T3 (= T5 for 100% effectiveness). Super-Calculate updates all the variables.

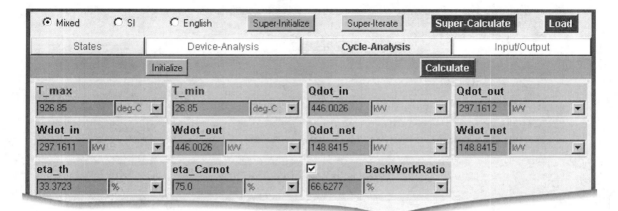

☐ **The updated efficiency can be found in the Cycle panel or in the I/O panel.**

■ *Gas Turbine Ex. #1*: Parametric Study–Impact of Regenerator Effectiveness

A Brayton cycle with regeneration. . . The thermal efficiency increases by about 5% as the effectiveness of the regenerator is increased from 65% to 100%. In a similar manner, the effect of compressor or turbine efficiency can be studied.

```
##################################################################
# To regenerate this solution, copy the following TEST-Code onto the I/O panel of the
# ..Open.Steady.Specific.PowerCycle.PhaseChanger daemon
# and click the Load and Super-Calculate buttons.
# ..***** A Super-Iterate beyond the Super-Calculate is  necessary in this problem*******
#--------------------Start of TEST-Code-------------------------------------------------------------------
----

States {
State-1:  Air;
Given:       { p1= 100.0 kPa;  T1= 300.0 K;  Vel1= 0.0 m/s;  z1= 0.0 m;  mdot1= 1.0 kg/s;  }

State-2:  Air;
Given:       { p2= 800.0 kPa;  Vel2= 0.0 m/s;  z2= 0.0 m;  j2= "j1+(j7−j1)/.82" kJ/kg;  mdot2= "mdot1" kg/s;  }

State-3:  Air;
Given:       { p3= "p2" kPa;  T3= "T2+0.65*(T5-T2)" K;  Vel3= 0.0 m/s;  z3= 0.0 m;  mdot3= "mdot1" kg/s;  }

State-4:  Air;
Given:       { p4= "p3" kPa;  T4= 1200.0 K;  Vel4= 0.0 m/s;  z4= 0.0 m;  mdot4= "mdot1" kg/s;  }

State-5:  Air;
Given:       {p5= 100.0 kPa;  Vel5= 0.0 m/s;  z5= 0.0 m;  j5= "j4−(j4−j8)*0.8" kJ/kg;  mdot5= "mdot1" kg/s;  }

State-6:  Air;
Given:       {p6= 100.0 kPa;  Vel6= 0.0 m/s;  z6= 0.0 m;  mdot6="mdot1"kg/s;  }

State-7:  Air;
Given:       {p7= 800.0 kPa;  s7= "s1" kJ/kg.K;  Vel7= 0.0 m/s;  z7= 0.0 m;  }

State-8:  Air;
Given:       {p8= 100.0 kPa;  s8= "s4" kJ/kg.K;  Vel8= 0.0 m/s;  z8= 0.0 m;  }
}

Analysis {
Device-A: i-State = State-1;  e-State = State-2;  Mixing: true
Given:  { Qdot= 0.0 kW;   T_B= 25.0 deg-C;  }

Device-B: i-State = State-3;  e-State = State-4;  Mixing: true
Given:  { Wdot_O= 0.0 kW;  T_B= 25.0 deg-C;  }

Device-C: i-State = State-4;  e-State = State-5;  Mixing: true
Given:  { Qdot= 0.0 kW;   T_B= 25.0 deg-C;  }

Device-D: i-State = State-2, State-5;  e-State = State-3, State-6;  Mixing: false
Given:  { Qdot= 0.0 kW;  Wdot_O= 0.0 kW;  T_B= 25.0 deg-C;  }

Device-E: i-State = State-6;  e-State = State-1;  Mixing: true
Given:  { Wdot_O= 0.0 kW;  T_B= 25.0 deg-C;  }

}
```

■ *Gas Turbine Ex. #1*: **TEST-Code to Regenerate the Visual Solution**

Open Cycles

Power plants (gas turbines and steam power) and refrigeration systems are modeled by the Brayton and Rankine cycles running forward or in reverse. Such cycles consist of open devices connected in a loop, and are called open cycles in TEST.

The Open Cycle daemons are built by adding a new layer—a cycle analysis panel for solving the cycle relations—to the Open Steady daemons. After the anchor states of the cycle are calculated, individual devices are analyzed. They are imported to the cycle panel, where the thermal efficiency and other relevant quantities are evaluated.

Example: A refrigerator uses R-12 as the working fluid and operates on an ideal vapor compression refrigeration cycle between 0.15 MPa and 1 MPa. If the mass flow rate is 0.04 kg/s, determine (a) the tonnage of the system, (b) the compressor power, and (c) the COP.

How would the answers change if the mass flow rate were doubled?

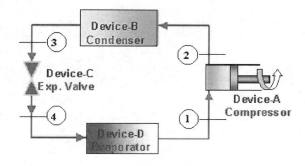

□ **Solution Procedure: Simplify the problem as an open, specific, refrigeration problem. Evaluate all the four states as best as possible. Analyze each device individually. Evaluate the unknowns with Super-Calculate. Change any parameter and Super-Calculate to update the solution.**

■ *Refrigeration Ex. #1*: **Problem Description**

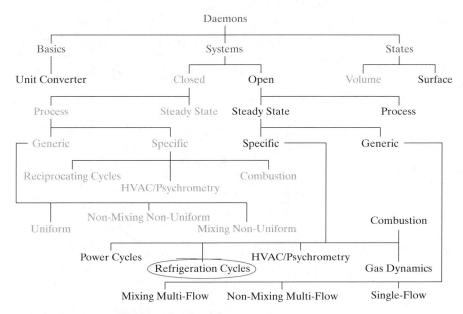

■ *Refrigeration Ex. #1*: **Simplify the Problem**

A refrigerator uses R-12. . . Use the TEST-Map to reach the appropriate daemon page: . . . Open.Steady.Specific.Refrigeration.

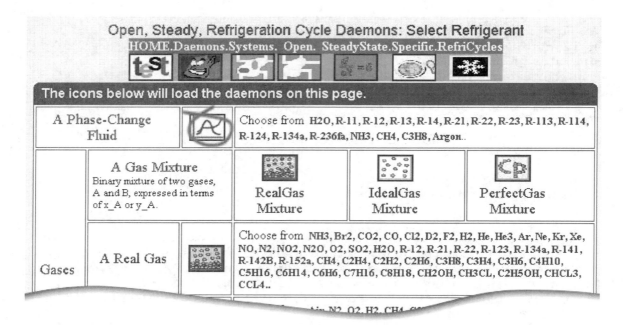

Open, Steady, Refrigeration Cycle Daemons: Select Refrigerant
HOME.Daemons.Systems. Open. SteadyState.Specific.RefriCycles

■ **Refrigeration Ex. #1: Choose a Working Fluid Model**

A refrigerator uses R-12... Choose the phase-change model, an obvious choice for R-12.

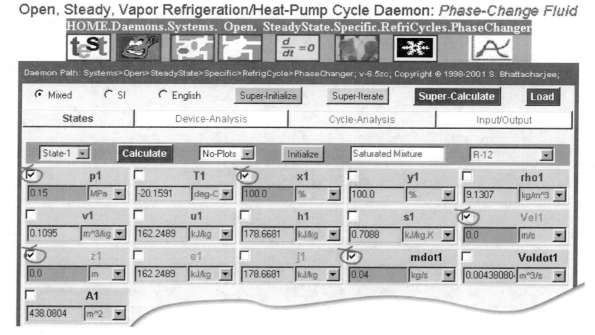

Open, Steady, Vapor Refrigeration/Heat-Pump Cycle Daemon: *Phase-Change Fluid*
HOME.Daemons.Systems. Open. SteadyState.Specific.RefriCycles.PhaseChanger

■ **Refrigeration Ex. #1: Evaluate the Compressor Inlet State (State-1)**

Refrigeration cycle between 0.15 MPa and 1 MPa... Choose State-1. Enter the known variables - p1, x1, and mdot 1. Calculate.

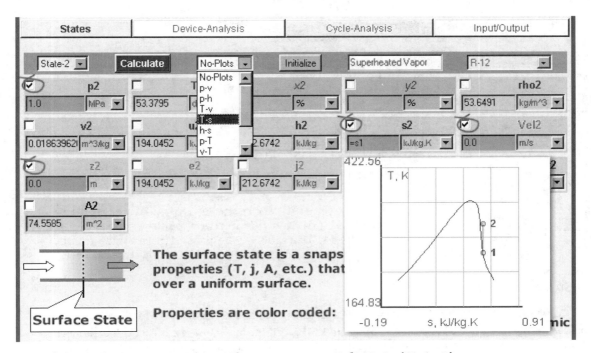

■ *Refrigeration Ex. #1*: Evaluate the Compressor Exit State (State-2)

Refrigeration cycle between 0.15 MPa and 1 MPa... If the mass flow rate is 0.04 kg/s, determine... Choose State-2. Enter known variables, p2, s2, mdot2 (optional), and Calculate.

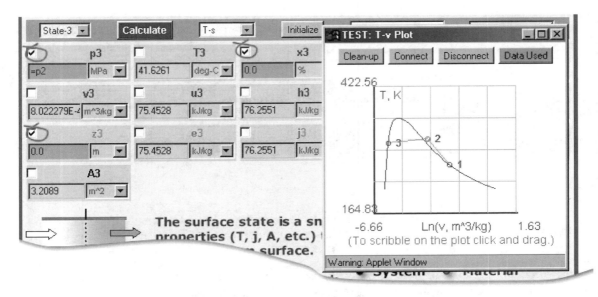

■ *Refrigeration Ex. #1*: Evaluate the Condenser Exit State (State-3)

An ideal refrigeration cycle between 0.15 MPa and 1 MPa... If the mass flow rate is 0.04 kg/s determine... Choose State-3. Enter known variables, p3, x3 and mdot3 (optional) and Calculate.

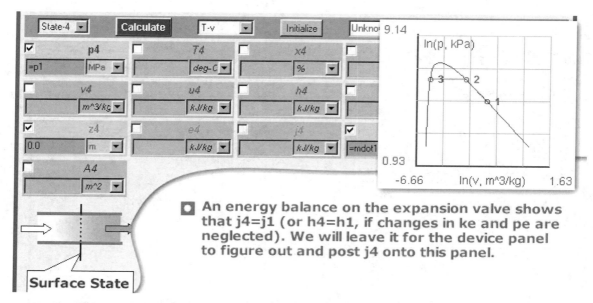

An energy balance on the expansion valve shows that j4=j1 (or h4=h1, if changes in ke and pe are neglected). We will leave it for the device panel to figure out and post j4 onto this panel.

Surface State

▪ *Refrigeration Ex. #1: Evaluate the Expansion Valve Exit State (State-4)*

An ideal refrigeration cycle between 0.15 MPa and 1 MPa... If the mass flow rate is 0.04 kg/s determine... Choose State-4. Enter known variables, p4 and mdot3 (optional) and Calculate the state partially opting for the Analysis panel to figure out and post j4 = j3.

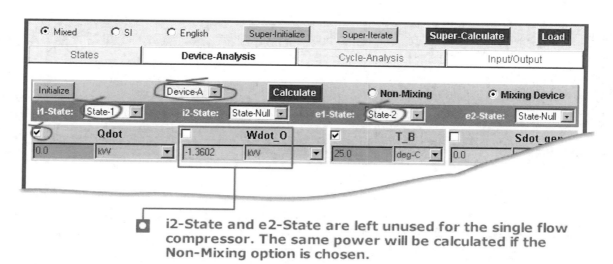

i2-State and e2-State are left unused for the single flow compressor. The same power will be calculated if the Non-Mixing option is chosen.

▪ *Refrigeration Ex. #1: Analyze the Compressor (Device-A)*

An ideal refrigeration cycle between 0.15 MPa and 1 MPa... If the mass flow rate is 0.04 kg/s determine .. Choose Device-A, load State-1 and state-2 as the inlet and exit states, enter Qdot Calculate.

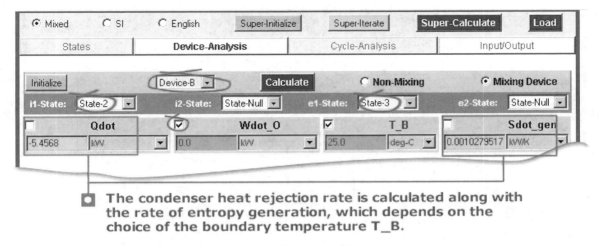

○ The condenser heat rejection rate is calculated along with the rate of entropy generation, which depends on the choice of the boundary temperature T_B.

■ *Refrigeration Ex. #1*: Analyze the Condenser (Device-B)

An ideal refrigeration cycle between 0.15 MPa and 1 MPa... If the mass flow rate is 0.04 kg/s determine .. Choose Device-B, load State-2 and State-3 as the inlet and exit states, enter Wdot and Calculate.

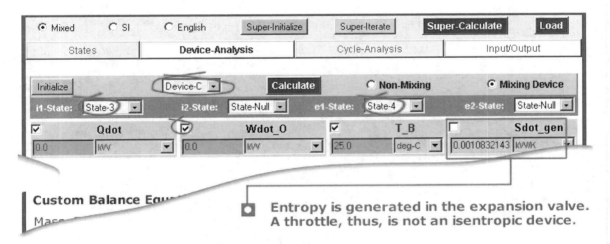

○ Entropy is generated in the expansion valve. A throttle, thus, is not an isentropic device.

■ *Refrigeration Ex. #1*: Analyze the Expansion Valve (Device-C)

An ideal refrigeration cycle between 0.15 MPa and 1 MPa... If the mass flow rate is 0.04 kg/s determine... Choose Device-C, load State-3 and State-4 as the inlet and exit states, enter Qdot Wdot and Calculate to find j4 = j3 which is automatically reported back to State-4.

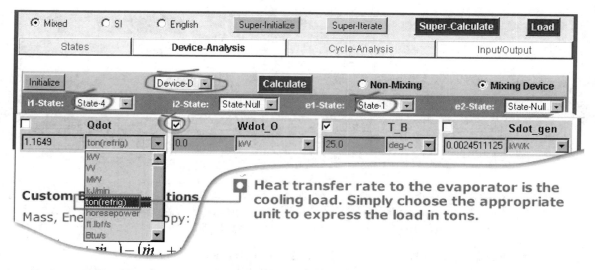

Heat transfer rate to the evaporator is the cooling load. Simply choose the appropriate unit to express the load in tons.

■ *Refrigeration Ex. #1*: Analyze the Evaporator (Device-D)

An ideal refrigeration cycle between 0.15 MPa and 1 MPa... If the mass flow rate is 0.04 kg/s determine .. Choose Device-D, load State-4 and State-1 as the inlet and exit States, enter Wdot and Calculate.

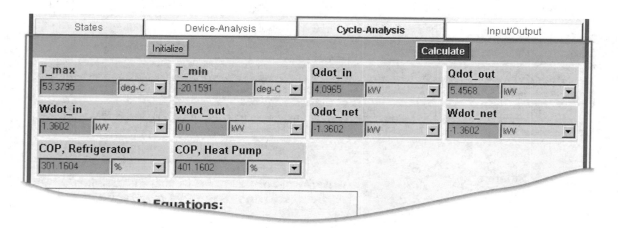

Devices analyzed must form a closed loop before the cycle analysis can be performed.

■ *Refrigeration Ex. #1*: Analyze the Cycle

An ideal refrigeration cycle between 0.15 MPa and 1 MPa... If the mass flow rate is 0.04 kg/s determine... All the devices having been analyzed, the Cycle panel is now ready for a Calculate.

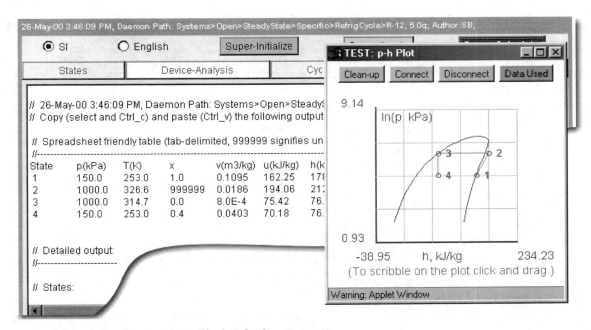

■ *Refrigeration Ex. #1*: Detailed Solution Report

How would the answers change if the mass flow rate is doubled?... Printer friendly output can be found after any Super-Calculate on the I/O panel.

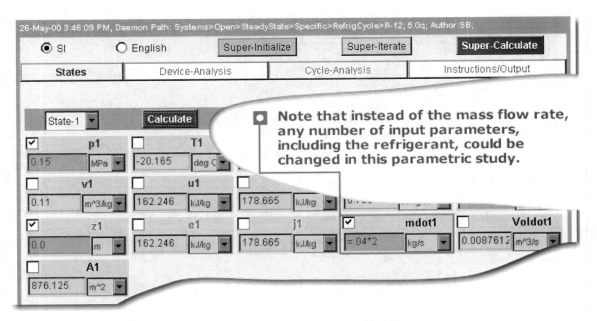

■ Refrigeration Examples: Parametric Study-Change the Flow Rate

How would the answers change if the mass flow rate were doubled?... Go to State-1 and change mdot1 to its new value. Calculate and Super-Calculate.

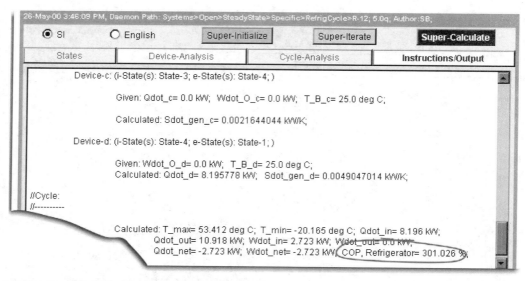

Refrigeration Ex. #1: Super-Calculate

How would the answers change if the mass flow rate were doubled? Switch to the Cycle or Output panel to find that the COP remains independent of the mass flow rate.

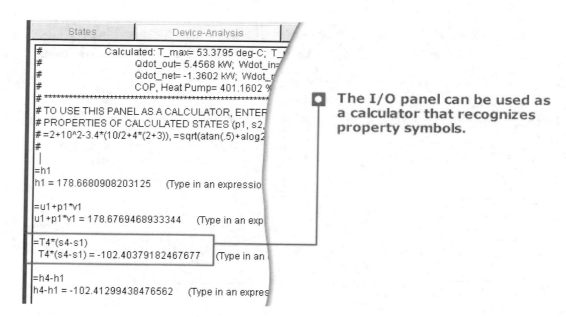

The I/O panel can be used as a calculator that recognizes property symbols.

Refrigeration Ex. #1: I/O Panel as a Calculator

An ideal refrigeration cycle between 0.15 MPa and 1 MPa... As with all other daemons, the I/O panel can be used as a scientific calculator.

```
####################################################################
# To regenerate this solution, copy the following TEST-Code onto the I/O panel of the
# ..Closed.Process.Specific.Refrigeration.PhaseChange daemon
# and click the Load and Super-Calculate buttons.
#--------------------Start of TEST-Code-------------------------------------------------------
----

 States {
 State-1:  R-12;
 Given:        { p1= 0.15 MPa;  x1= 100.0 %;  Vel1= 0.0 m/s;  z1= 0.0 m;  mdot1= 0.04 kg/s;  }

 State-2:  R-12;
 Given:        { p2= 1.0 MPa;  s2= "s1" kJ/kg.K;  Vel2= 0.0 m/s;  z2= 0.0 m;
              mdot2= "mdot1" kg/s;  }

 State-3:  R-12;
 Given:        { p3= "p2" MPa;  x3= 0.0 %;  Vel3= 0.0 m/s;  z3= 0.0 m;
              mdot3= "mdot2" kg/s;  }

 State-4:  R-12;
 Given:        { p4= "p1" MPa;  Vel4= 0.0 m/s;  z4= 0.0 m;  mdot4= "mdot1" kg/s;  }
 }

 Analysis {
 Device-A: i-State = State-1;  e-State = State-2;  Mixing: true;
 Given: { Qdot= 0.0 kW;  T_B= 25.0 deg-C;  }

 Device-B: i-State = State-2;  e-State = State-3;  Mixing: true;
 Given: { Wdot_O= 0.0 kW;  T_B= 25.0 deg-C;  }

 Device-C: i-State = State-3;  e-State = State-4;  Mixing: true;
 Given: { Qdot= 0.0 kW;  Wdot_O= 0.0 kW;  T_B= 25.0 deg-C;  }

 Device-D: i-State = State-4;  e-State = State-1;  Mixing: true;
 Given: { Wdot_O= 0.0 kW;  T_B= 25.0 deg-C;  }

 }
```

■ *Refrigeration Ex. #1*: **TEST-Code to Regenerate the Visual Solution**

HVAC

The **HVAC** daemons, built upon the **Moist Air State** daemon, appear under Closed-Process, as well as the Open and Steady branches in the TEST-Map. A background in the Closed-Process (Chapter 3) and Open-Steady (Chapter 4) daemons is essential to understanding the workings of the HVAC daemons. The Analysis panel, in particular, is quite versatile. By using appropriate combinations of radio buttons and the inlet and exit states, a variety of psychrometric applications, such as simple heating, simple cooling, humidification, dehumidification, and wet cooling towers, can be studied.

The data range for moist air is quite high, extending the dew point temperature down to −40 deg-C. A word of caution about the units used for the moist air is in order: Because the same state interface is used for moist air, steam, and condensed water, units that involve mass should be properly interpreted; thus, "kg" for moist air should read "kg of dry air." In addition to the examples presented here, the Tutorial and Chapter 13 of the Archive contain a large number of examples with the HVAC daemons.

Example: Moist air at 40 deg C and 90% R.H. enters a dehumidifier at the rate of 300 m3/min. The condensate and the saturated air exit at 10 deg C through separate exits. The pressure remains constant at 100 kPa. Determine (a) the mass flow rate of dry air, (b) the water removal rate, and (c) the required refrigeration capacity, in tons. How would the answers change if the inlet humidity were 20% instead?

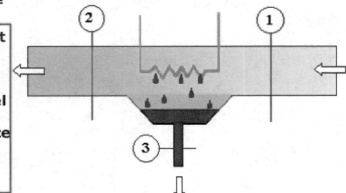

> ◙ Solution Procedure: Start the Open HVAC daemon. Evaluate the inlet and exit states as best as possible. Configure the analysis panel for a single inlet and two exits. Solution of the balance equations produces state variables. Super-Calculate finds the desired answers.

■ *HVAC Ex. #1*: **Problem Description**

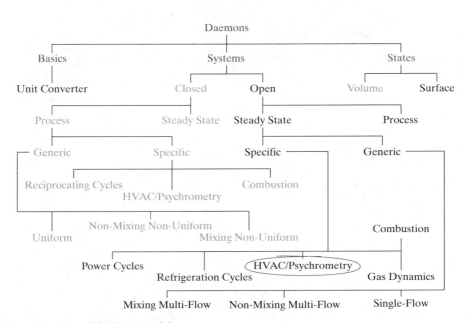

■ *HVAC Ex. #1*: **Simplify the Problem**

Moist air at 40 deg-C and 90% R.H ... The appropriate daemon for this problem is ...
Open.Steady.Specific.HVAC.

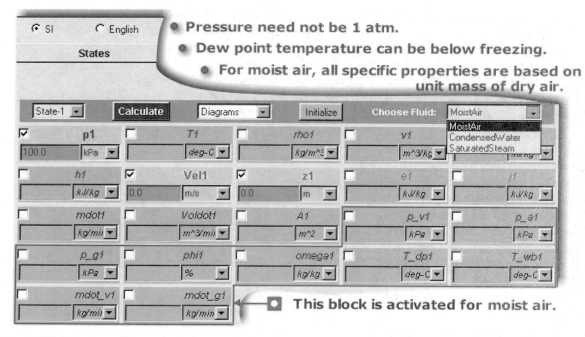

■ *HVAC Ex. #1:* The Moist Air State Panel

Moist air at 40 deg-C and 90% R.H. ... The state panel for moist air contains a few extra variables (boxed in the red polygon), but exhibits the same look and feel as any other state panel.

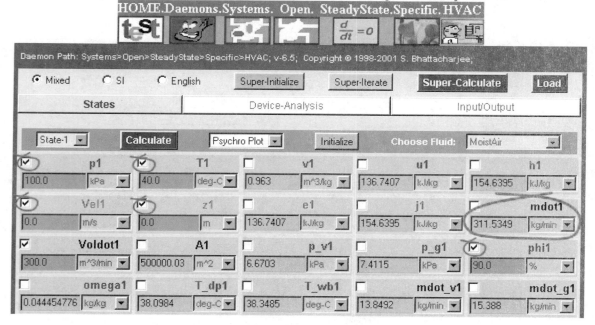

■ *HVAC Ex. #1:* Evaluate the Moist Air Inlet State (State-1)

Moist air at 40 deg-C and 90% R.H. ... Enter p1, T1, Voldot1, and phi1. Calculate produces mdot1, the flow rate of dry air (311 kg of dry air/min).

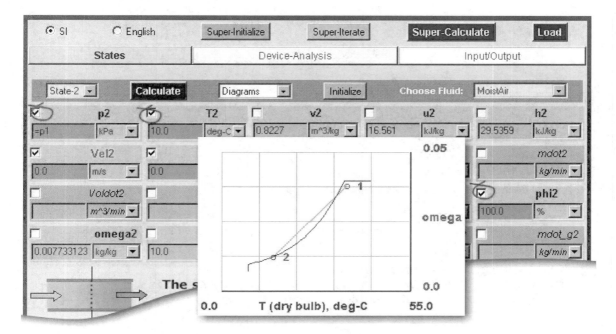

■ *HVAC Ex. #1*: Evaluate the Moist Air Exit State (State-2)

Moist air at 40 deg-C and 90% R.H. . . . Enter p2, T2, and phi2. Calculate. Although the psychrometric plot joins state-1 and state-2 with a straight line, the actual path, which is an unknown, cannot cross the saturation

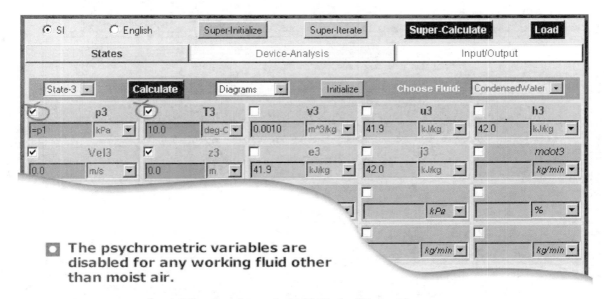

□ **The psychrometric variables are disabled for any working fluid other than moist air.**

■ *HVAC Ex. #1*: Evaluate the Condensate Exit State (State-3)

Moist air at 40 deg-C and 90% R.H. . . . Choose Condensed Water from the working fluid selector, enter p3 and T3. Calculate.

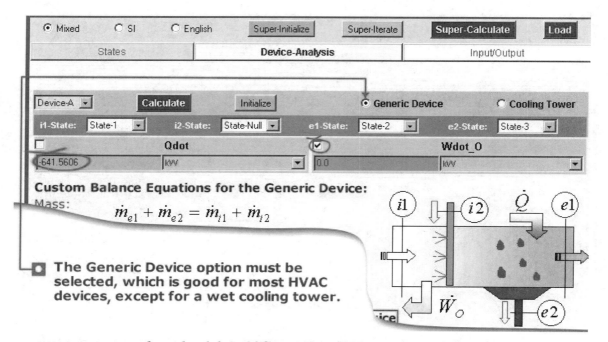

Custom Balance Equations for the Generic Device:

Mass:

$$\dot{m}_{e1} + \dot{m}_{e2} = \dot{m}_{i1} + \dot{m}_{i2}$$

The Generic Device option must be selected, which is good for most HVAC devices, except for a wet cooling tower.

■ *HVAC Ex. #1*: Analyze the dehumidifier—Solve the Balance Equations

Moist air at 40 deg-C and 90% R.H. . . . Enter Wdot_O, and Calculate. Super-Calculate produces a cooling load of 641.5 kW and posts the condensate flow rate in state-3.

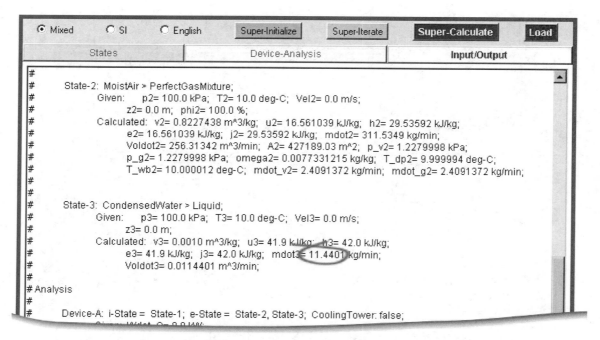

■ *HVAC Ex. #1*: The Detailed Report

Moist air at 40 deg-C and 90% R.H. . . . The detailed report on the I/O panel also contains the condensate flow rate.

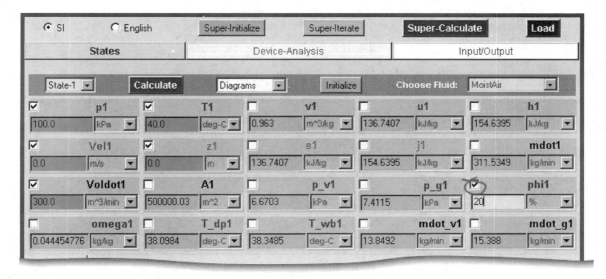

■ HVAC Ex. #1: Parametric Study—Change Relative Humidity

Moist air at 40 deg-C and 90% R.H. ... Go back to state-1 and change phi1 to 20%. Press the Enter key or the Calculate button to register the change. Super-Calculate updates all the dependent variables.

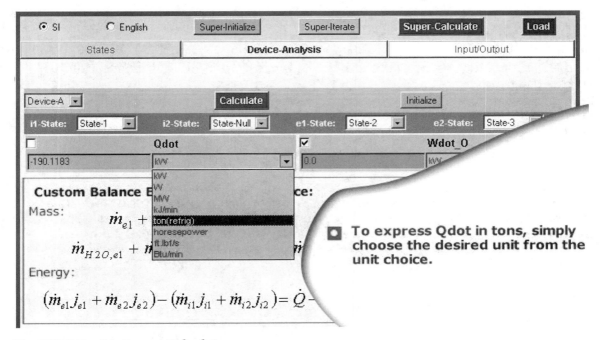

To express Qdot in tons, simply choose the desired unit from the unit choice.

■ HVAC Ex. #1: Super-Calculate

Moist air at 40 deg-C and 90% R.H. ... The new cooling load is significantly smaller, due to the fact that much less moisture needs to be condensed. It is cheaper to cool dry air.

```
###################################################################
# To regenerate this solution, copy the following TEST-Code onto the I/O panel of the
# ..Open.Steady.Specific.HVAC daemon
# and click the Load and Super-Calculate buttons.
#-------------------Start of TEST-Code----------------------------------------------------------------
----

  States {
  State-1:  MoistAir;
  Given:      { p1= 100.0 kPa;  T1= 40.0 deg-C;  Vel1= 0.0 m/s;  z1= 0.0 m;
                 Voldot1= 300.0 m^3/min;  phi1= 90.0 %;  }

  State-2:  MoistAir;
  Given:      { p2= 100.0 kPa;  T2= 10.0 deg-C;  Vel2= 0.0 m/s;  z2= 0.0 m;  phi2= 100.0
  %;  }

  State-3:  CondensedWater;
  Given:      { p3= 100.0 kPa;  T3= 10.0 deg-C;  Vel3= 0.0 m/s;  z3= 0.0m;  }
  }

  Analysis {
  Device-A: i-State = State-1;  e-State = State-2, State-3;  CoolingTower: false;
  Given: { Wdot_O= 0.0 kW;  }
  }
```

■ **HVAC Ex. #1: TEST-Code to Regenerate the Visual Solution**

Example: A 12'x12'x10' insulated chamber has air at 1 atm, 70 deg-F, and 90% relative humidity (R.H.). Determine the final pressure, temperature, and R. H., if 1 kW.Hr of electrical work is transferred to the chamber.

How would these answers change if the initial R.H. were 5%?

◻ **Solution Procedure: Simplify the problem as a closed, specific, HVAC problem. Enter the known information about the begin and finish states. Analyze the process and Super-Calculate to find the desired answers.**

■ **HVAC Ex. #2: Problem Description**

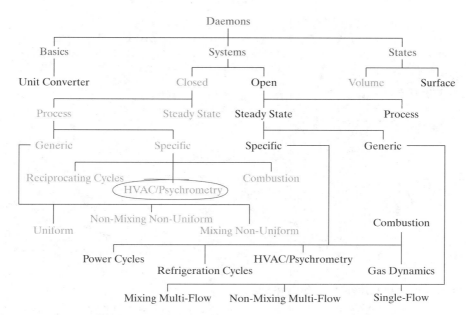

■ *HVAC Ex. #2*: **Simplify the Problem**

A 12′ × 12′ × 10′ chamber at 1 atm ... The appropriate classification for this problem is ... Closed.Process.Specific.HVAC.

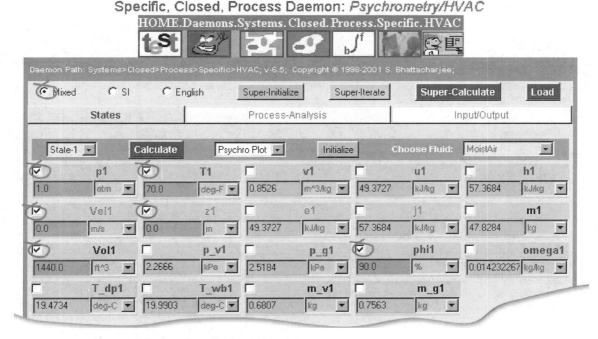

■ *HVAC Ex. #2*: **Evaluate the Begin-State**

A 12′ × 12′ × 10′ chamber at 1 atm ... Let us work with mixed units in this example. Enter p1, T1, phi1 and Vol1 (note the use of equation, Vol1 = 12*12*10) in appropriate units. Calculate.

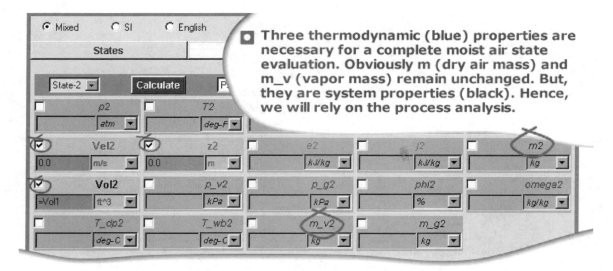

Three thermodynamic (blue) properties are necessary for a complete moist air state evaluation. Obviously m (dry air mass) and m_v (vapor mass) remain unchanged. But, they are system properties (black). Hence, we will rely on the process analysis.

■ HVAC Examples: Prepare the Finish-State

A 12′ × 12′ × 10′ chamber at 1 atm ... Select state-2 as the f-state. The only known information in the volume does not change. The default pressure of 100 kPa must be unchecked.

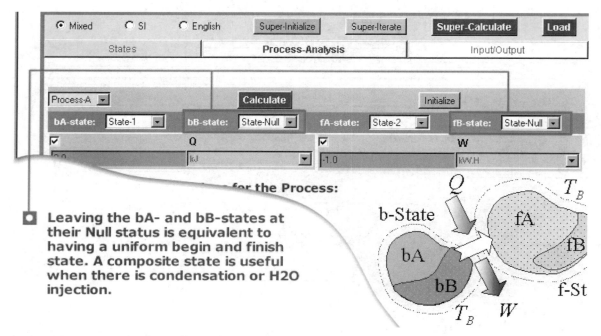

Leaving the bA- and bB-states at their Null status is equivalent to having a uniform begin and finish state. A composite state is useful when there is condensation or H2O injection.

■ *HVAC Ex. #2: Analyze the Process*

A 12′ × 12′ × 10′ chamber at 1 atm ... Load the b- and f-states. Although the daemon allows for composite states, we will use only one sub-system, say, A. Enter W and Q. Super-Calculate.

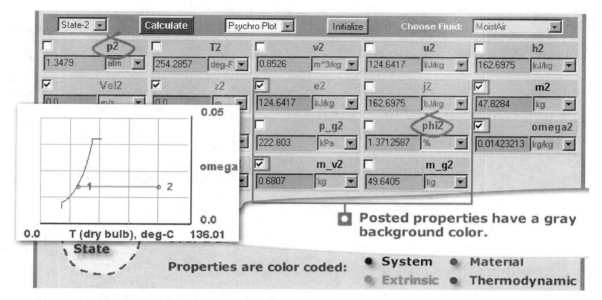

Posted properties have a gray background color.

Properties are color coded: ● **System** ● **Material**
● Extrinsic ● **Thermodynamic**

■ *HVAC Ex. #2:* **State-2 Evaluated**

A 12′ × 12′ × 10′ chamber at 1 atm . . . The total pressure is found. The analysis panel posts m, m omega and e. With Vel and z set to zero, two thermodynamic properties u and omega have been found.

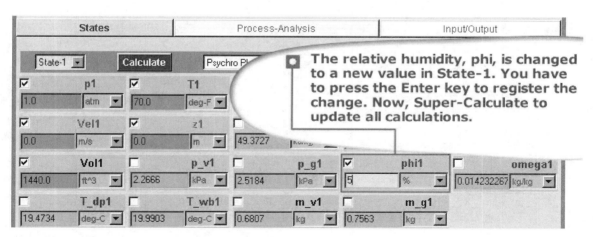

The relative humidity, phi, is changed to a new value in State-1. You have to press the Enter key to register the change. Now, Super-Calculate to update all calculations.

■ *HVAC Ex. #2:* **Change an Input Parameter**

A 12′ × 12′ × 10′ chamber at 1 atm . . . The R.H. at State-1 is changed to a very small value, say, 1.0, to see the effect of humidity on the final pressure and temperature.

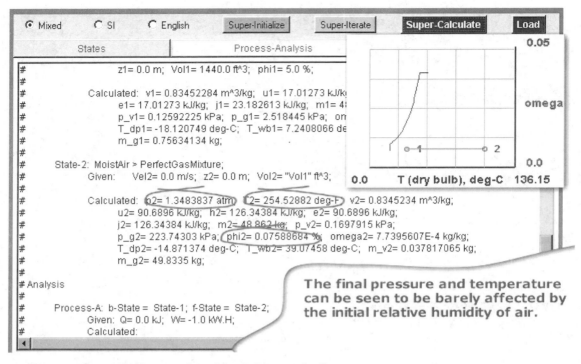

■ *HVAC Ex. #2:* **Solution Updated by Super-Calculate**

A 12′ × 12′ × 10′ chamber at 1 atm ... Super-Calculate shifts the focus to the I/O panel, where the solution report and the TEST-Code are generated.

```
####################################################################
# To regenerate this solution, copy the following TEST-Code onto the I/O panel of the
# ..Closed.Process.Specific.HVAC daemon
# and click the Load and Super-Calculate buttons.
#--------------------Start of TEST-Code----------------------------------------------------------------
----

 States {
 State-1: MoistAir;
 Given:     { p1= 1.0 atm;  T1= 70.0 deg-F;  Vel1= 0.0 m/s;  z1= 0.0 m;  Vol1= 1440.0
ft^3;  phi1= 90.0 %;  }

 State-2: MoistAir;
 Given:     { Vel2= 0.0 m/s;  z2= 0.0 m;  Vol2= "Vol1" ft^3;  }
 }

 Analysis {
 Process-A: b-State = State-1;  f-State = State-2;
 Given: { Q= 0.0 kJ;  W= −1.0 kW.H;  }
 }
```

■ *HVAC Ex. #2:* **TEST-Code to Regenerate Visual Solution**

Example: Warm water leaves a processing plant at 35 deg C at a rate of 50 kg/s. The water is cooled to 20 deg C in a cooling tower by air that enters the tower at 1 bar, 20 deg C, and 60% relative humidity and leaves saturated at 30 deg C. Determine (a) the volume flow rate of air and (b) the rate of loss of water in the tower. Neglect any power input to the induced-draft fan or heat transfer.

How would the answers change if the relative humidity dropped to 20%?

> **▣Solution Procedure: Simplify the problem as an open, specific, HVAC problem. Evaluate the air inlet and exit states and the water inlet and exit states as best as possible. The balance equation produces the air mass flow rate. Super-Calculate completes all calculations.**

■ *HVAC Ex. #3*: **Problem Description**

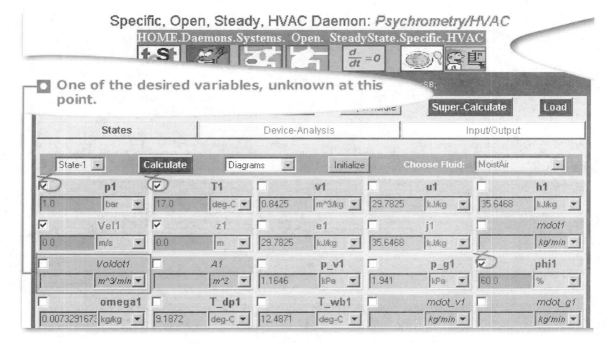

■ *HVAC Ex. #3*: **Evaluate the Air Inlet State (State-1)**

The water is cooled to 20 deg-C in a cooling tower ... Start the ..Open.Steady.HVAC daemon. For State-1, select Moist Air as the working fluid, enter p1, T1, and phi1, and Calculate.

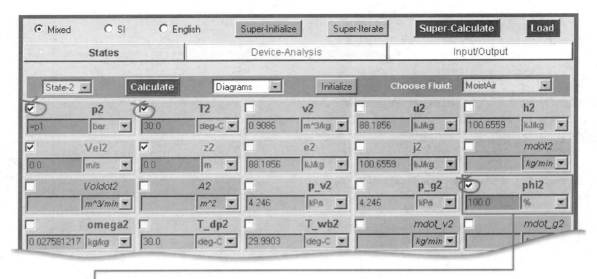

☐ The air exit state, State-2, is saturated. That is, the relative humidity is 100%.

■ *HVAC Ex. #3:* **Evaluate the Air Exit State (State-2)**

The water is cooled to 20 deg-C in a cooling tower ... Select State-2, enter p2, T2, and phi2. Calculate.

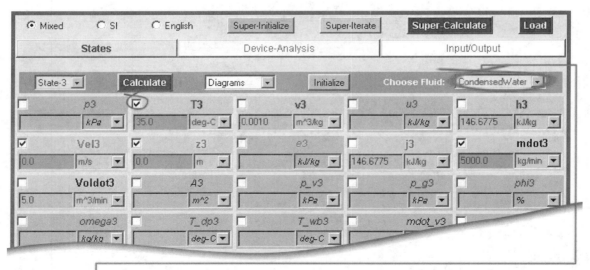

☐ The water temperature is sufficient for evaluating its flow energy j. Only mdot3 and j3 are used in the energy equation.

■ *HVAC Ex. #3:* **Evaluate the Water Inlet State (State-3)**

The water is cooled to 20 deg-C in a cooling tower ... Select State-3, choose Condensed Water to model liquid water, enter T3, and Calculate.

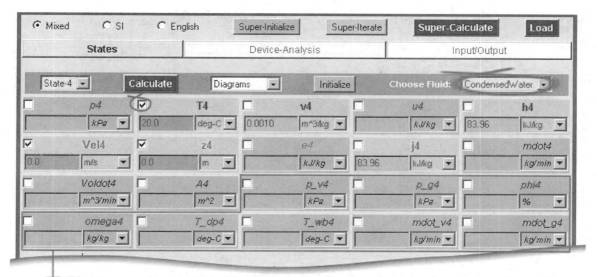

■ The properties in this block are used only when moist air is selected as the working fluid. The mass flow rate, mdot4, is one of the desired unknowns with the water loss rate being mdot4-mdot3.

■ *HVAC Ex. #3*: Evaluate the Water Exit State (State-4)

The water is cooled to 20 deg-C in a cooling tower ... Select State-4, choose Condensed Water to model liquid water, enter T4, and Calculate.

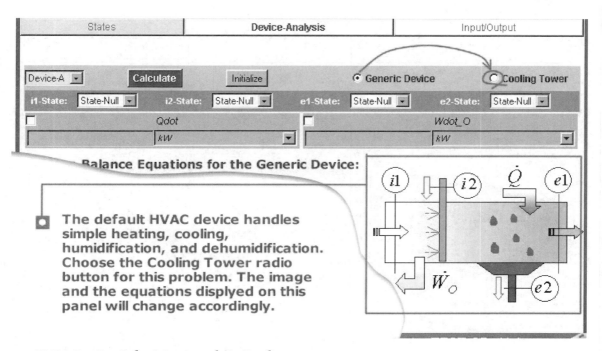

■ The default HVAC device handles simple heating, cooling, humidification, and dehumidification. Choose the Cooling Tower radio button for this problem. The image and the equations displyed on this panel will change accordingly.

■ *HVAC Ex. #3*: Select Appropriate Device

The water is cooled to 20 deg-C in a cooling tower ... Select the Cooling Tower option in the Analysis panel. The system schematic and governing equations change accordingly.

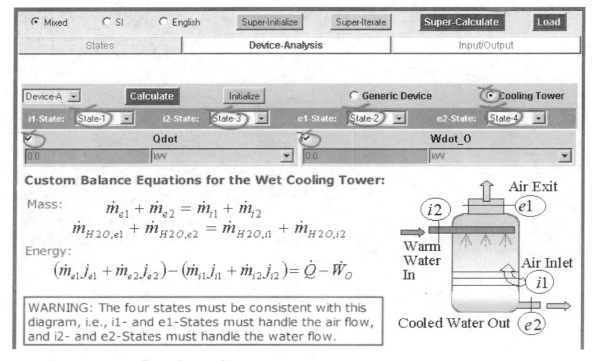

HVAC Ex. #3: Analyze the Cooling Tower

The water is cooled to 20 deg-C in a cooling tower ... Load the inlet and exit states in a manner consistent with the schematic. Enter Qdot and Wdot_O, Calculate, and Super-Calculate. The mass flow rate of air is evaluated and posted back to State-1.

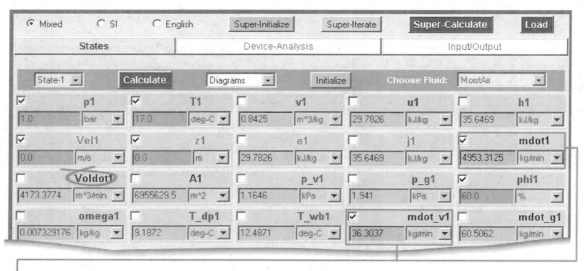

> **⬛** The flow rate of dry air and water vapor, mdot and mdot_v, are posted by the analysis panel (note the gray background), and State-1 is completely evaluated, including the desired unknown Voldot, by Super-Calculate.

HVAC Ex. #3: Volume Flow Rate of Air

The water is cooled to 20 deg-C in a cooling tower ... Note that the volume flow rate of air at the inlet (State-1) is different (smaller) than the volume flow rate at the exit (State-2).

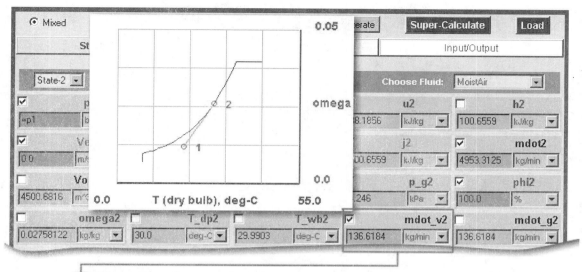

> The rate of water loss is the same as the rate of moisture gain by the air: mdot_v2-mdot_v1= 136.6-36.3=100.3 kg/min.

■ *HVAC Ex. #3:* Rate of Water Loss

The water is cooled to 20 deg-C in a cooling tower ... The water loss can also be calculated by finding the difference in water flow rate (mdot) between State-3 and State-4.

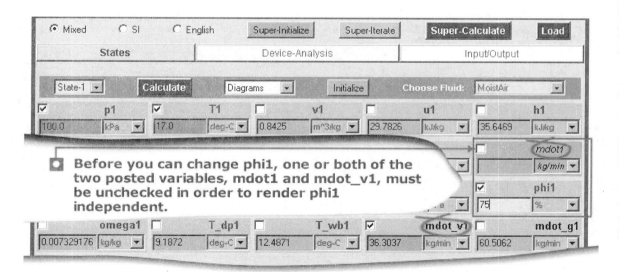

> Before you can change phi1, one or both of the two posted variables, mdot1 and mdot_v1, must be unchecked in order to render phi1 independent.

> After phi1 is changed, press the Enter key (or the Calculate button) to register the change, and Super-Calculate to update all variables in all panels.

■ *HVAC Ex. #3:* Parametric Study—Effect of Relative Humidity

The water is cooled to 20 deg-C in a cooling tower ... Go back to State-1 and change phi1 (R.H.) to a new value, 75%.

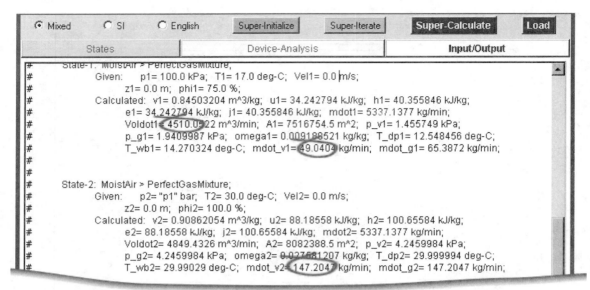

```
 ○ Mixed      ○ SI      ○ English      Super-Initialize      Super-Iterate      Super-Calculate      Load
            States                    Device-Analysis                    Input/Output
# State-1: MoistAir > PerfectGasMixture;
#         Given:    p1= 100.0 kPa;  T1= 17.0 deg-C;  Vel1= 0.0 m/s;
#                   z1= 0.0 m;  phi1= 75.0 %;
#         Calculated:  v1= 0.84503204 m^3/kg;  u1= 34.242794 kJ/kg;  h1= 40.355846 kJ/kg;
#                   e1= 34.242794 kJ/kg;  j1= 40.355846 kJ/kg;  mdot1= 5337.1377 kg/min;
#                   Voldot1= 4510.0522 m^3/min;  A1= 7516754.5 m^2;  p_v1= 1.455749 kPa;
#                   p_g1= 1.9409987 kPa;  omega1= 0.009188521 kg/kg;  T_dp1= 12.548456 deg-C;
#                   T_wb1= 14.270324 deg-C;  mdot_v1= 49.0404 kg/min;  mdot_g1= 65.3872 kg/min;
#
#
#
# State-2: MoistAir > PerfectGasMixture;
#         Given:    p2= "p1" bar;  T2= 30.0 deg-C;  Vel2= 0.0 m/s;
#                   z2= 0.0 m;  phi2= 100.0 %;
#         Calculated:  v2= 0.90862054 m^3/kg;  u2= 88.18558 kJ/kg;  h2= 100.65584 kJ/kg;
#                   e2= 88.18558 kJ/kg;  j2= 100.65584 kJ/kg;  mdot2= 5337.1377 kg/min;
#                   Voldot2= 4849.4326 m^3/min;  A2= 8082388.5 m^2;  p_v2= 4.2459984 kPa;
#                   p_g2= 4.2459984 kPa;  omega2= 0.027581207 kg/kg;  T_dp2= 29.999994 deg-C;
#                   T_wb2= 29.99029 deg-C;  mdot_v2= 147.2047 kg/min;  mdot_g2= 147.2047 kg/min;
```

▣ **The water loss (147-49=98 kg/min) hardly changes (why?), while the volume flow rate goes up significantly.**

■ *HVAC Ex. #3: Update All Calculations—Super-Calculate*

The water is cooled to 20 deg-C in a cooling tower ... The solution report generated by Super-Calculate in the I/O panel displays the new results, as do the individual States and Analysis panels.

```
#####################################################################
# To regenerate this solution, copy the following TEST-Code onto the I/O panel of the
# ..Open.Steady.Specific.HVAC daemon
# and click the Load and Super-Calculate buttons.
#-------------------Start of TEST-Code------------------------------------------------------
----

States {
State-1:  MoistAir;
Given:       { p1= 100.0 kPa;  T1= 17.0 deg-C;  Vel1= 0.0 m/s;  z1= 0.0 m;  phi1= 60.0 %;  }

State-2:  MoistAir;
Given:       { p2= "p1" bar;  T2= 30.0 deg-C;  Vel2= 0.0 m/s;  z2= 0.0 m;  phi2= 100.0 %;  }

State-3:  CondensedWater;
Given:       { T3= 35.0 deg-C;  Vel3= 0.0 m/s;  z3= 0.0 m;  mdot3= 5000.0 kg/min;  }

State-4:  CondensedWater;
Given:       {T4= 20.0 deg-C;  Vel4= 0.0 m/s;  z4= 0.0 m;  }
}

Analysis {
Device-A: i-State = State-1, State-3;  e-State = State-2, State-4;  CoolingTower: true;
Given: { Qdot= 0.0 kW;  Wdot_O= 0.0 kW;  }
}
```

■ *HVAC Ex. #3: TEST-Code to Regenerate the Visual Solution*

Example: A 12'x12'x10' chamber has air at 1 atm, 70 deg-C and 30% relative humidity. Determine the heat transfer and the amount of H2O necessary to saturate the air (a) by injecting saturated steam at 1 atm, (b) by spraying liquid water at 70 deg-F, (c) by cooling the room at constant pressure (1 atm),(d) by spraying liquid water while not allowing any heat transfer to the chamber.

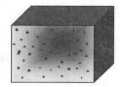

Show these saturation processes on a psychrometric plot. How would the plot change if the chamber temperature were 100 deg-F instead?

> 🔲 **Solution Procedure: Simplify the problem as a closed, specific, HVAC problem. Evaluate the composite begin states and the saturated finish state. Analyze the saturation process and Super-Calculate to find the desired answers. Repeat for each process.**

▪ *HVAC Ex. #4:* **Problem Description**

Example: A $12' \times 12' \times 10'$ chamber has air at 1 atm, 70 deg-C and 30% relative humidity. Determine the heat transfer and the amount of H20 necessary to saturate the air (a) by injecting saturated steam at 1 atm, (b) by spraying liquid water at 70 deg-F, (c) by cooling the room at constant pressure (1 atm), (d) by spraying liquid water while not allowing any heat transfer to the chamber.

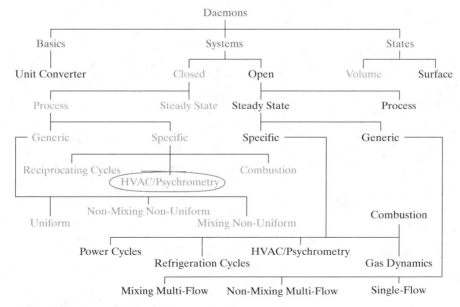

▪ *HVAC Ex. #4:* **Simplify the Problem**

A 12' × 12' × 10' chamber at 1 atm... The appropriate classification for this problem is ... Closed.Process.Specific.HVAC.

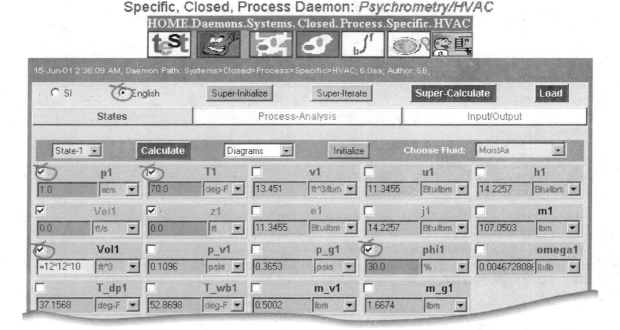

HVAC Ex. #4: Evaluate the Composite Begin-State (State-1 for bA-State)

A 12′ × 12′ × 10′ chamber at 1 atm ... In the first part of the problem, the composite begin state is made of moist air and steam. Choose English units, enter p1, T1, phi1 and Vol1 (note the equation, Vol1 = 12*12*10). Calculate.

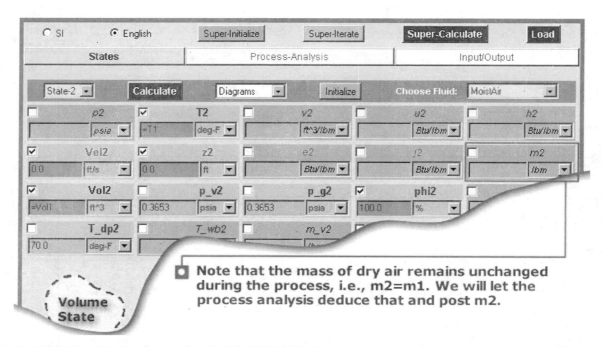

Note that the mass of dry air remains unchanged during the process, i.e., m2=m1. We will let the process analysis deduce that and post m2.

HVAC Ex. #4: Evaluate the Finish-State (State-2)

A 12′ × 12′ × 10′ chamber at 1 atm ... Select state-2 as the f-state for isothermal saturation. Enter Vol2 and phi2 (100% for a saturated state). Note that p2 should not be assumed equal to p1. Calculate.

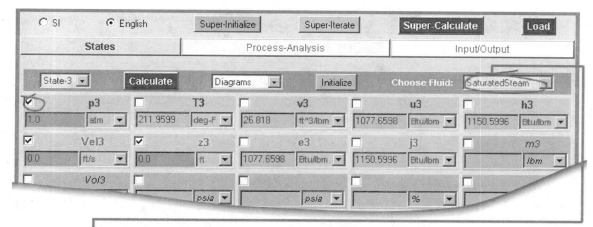

For saturated steam, the only information necessary is the pressure or temperature.

■ HVAC Ex. #4: Evaluate the Begin-State for Steam (State-3 for bB-State)

A 12′ × 12′ × 10′ chamber at 1 atm ... Select state-3 as the bB-state for saturated steam. Simply enter the p3 and Calculate.

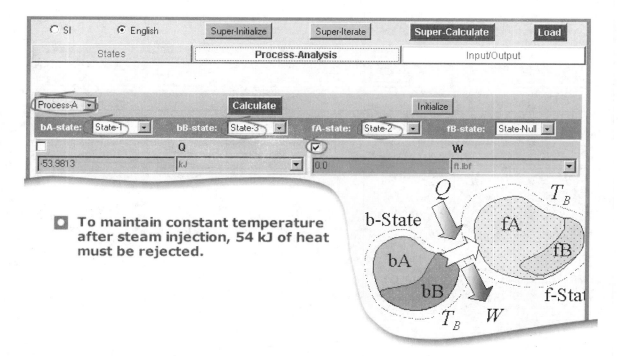

To maintain constant temperature after steam injection, 54 kJ of heat must be rejected.

■ *HVAC Ex. #4:* Analyze the Isothermal Saturation Process

A 12′ × 12′ × 10′ chamber at 1 atm ... Load the composite b-states and the single f-state. Enter W Super-Calculate. The solutions of the mass balance equations are posted back to state-2 and 3. Here we neglect the work done in injecting the steam.

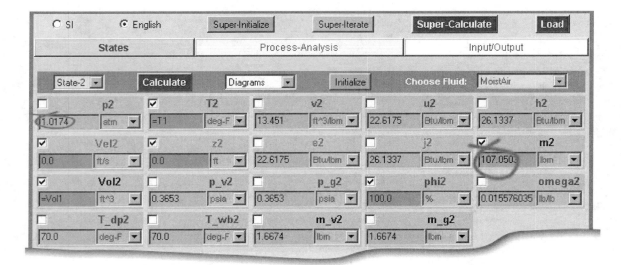

■ *HVAC Ex. #4*: State-2 Evaluated

A 12′ × 12′ × 10′ chamber at 1 atm ... The total pressure is slightly higher than 1 atm due to steam injection. The mass addition can be found from state-3 or from m_v2-m_v1.

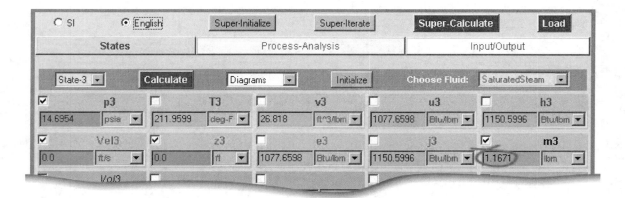

■ *HVAC Ex. #4*: State-3 Evaluated

A 12′ × 12′ × 10′ chamber at 1 atm ... The mass added, m3, is posted on State-3.

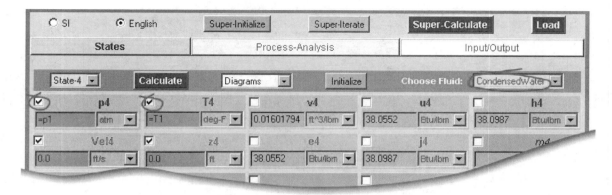

- The change in water pressure necessary to continue spraying in a closed atmosphere of slightly rising pressure (due to mass addition) is considered negligible.

HVAC Ex. #4: Evaluate the State of Water Spray (State-4)

A 12′ × 12′ × 10′ chamber at 1 atm ... For the saturation process with liquid water, choose Condensed Water for state-4. Enter p4 and T4. Calculate.

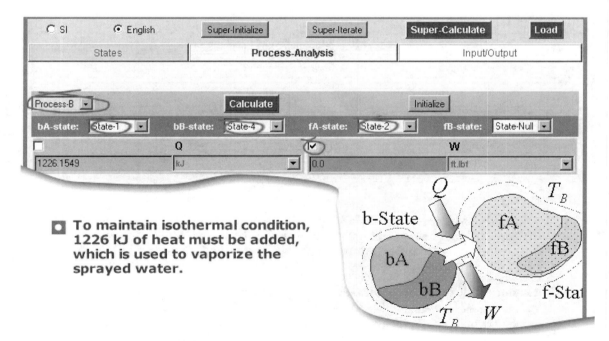

- To maintain isothermal condition, 1226 kJ of heat must be added, which is used to vaporize the sprayed water.

HVAC Ex. #4: Analyze the Isothermal Saturation Process

A 12′ × 12′ × 10′ chamber at 1 atm ... Select Process-B. Load the b- and f-states, enter W = 0. Super Calculate. All the panels are updated. Notice that a significant amount of heat must be added maintain the isothermal condition.

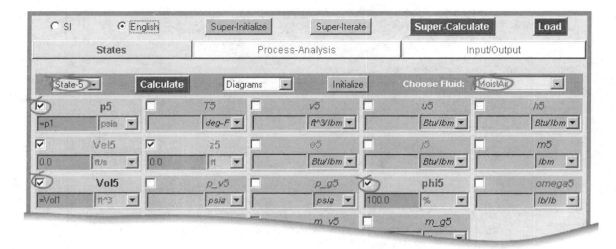

■ *HVAC Ex. #4:* **Saturated Air (State-5) From Constant Pressure Cooling**

A 2′ × 12′ × 10′ chamber at 1 atm ... For constant pressure cooling, select state-5 as the f-state Enter p5, Vol5 and phi5. To complete the state, the cooling process (State-1 to State-5) must be analyzed.

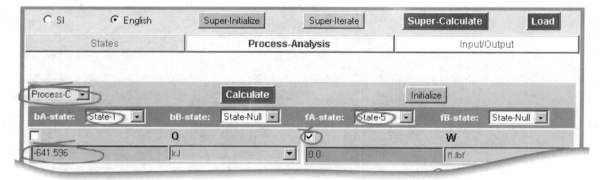

■ The Super-Calculate operation posts solutions from the balance equations for dry air mass and H2O mass into state-5. The energy balance equation produces a heat loss of 641.5 kJ to reach saturation at dew point.

■ *HVAC Ex. #4:* **Analyze the Constant-Pressure Cooling Process**

A 12′ × 12′ × 10′ chamber at 1 atm ... Select Process-C, load the b- and f-states. Enter W and Super-Calculate. Heat transfer Q is found and State-5 is updated.

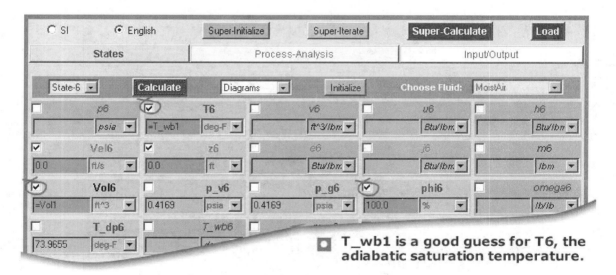

■ *HVAC Ex. #4*: Dew Point Saturation (State-5) Evaluated

A 12′ × 12′ × 10′ chamber at 1 atm ... Note that the temperature calculated, T5, is the dew point temperature of state-1 and state-5. A constant pressure cooling leads to saturation at the dew point.

☐ **T_wb1 is a good guess for T6, the adiabatic saturation temperature.**

■ *HVAC Ex. #4*: Evaluate the Adiabatic Saturation State (State-6)

A 12′ × 12′ × 10′ chamber at 1 atm ... To evaluate the adiabatic saturation state, use T_wb1 as the first guess for T6. Enter Vol6 and phi6, and Calculate.

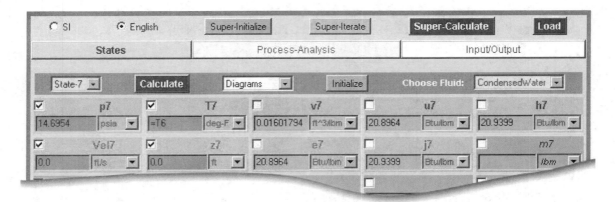

■ Moderate variation of the temperature of the condensed water added during saturation does not affect the final temperature much. A parametric study on the effect of T7 can be used to verify that.

■ *HVAC Ex. #4:* Evaluate the State of Water Spray

A 12′ × 12′ × 10′ chamber at 1 atm ... The water added for saturation is evaluated at the saturation temperature T6 following convention.

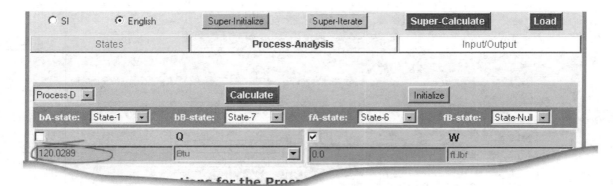

■ To determine the adiabatic saturation temperature, one has to iterate by guessing T6 and Super-Calculating Q until Q is sufficiently close to zero. (T6 will be smaller than the current solution by less than half a degree).

■ *HVAC Ex. #4:* Analyze the Adiabatic Saturation Process (Process-D)

A 12′ × 12′ × 10′ chamber at 1 atm ... Select Process-D, load the b- and f-states, enter W = 0. Super-Calculate to find Q. Now go back and change T6 and Super-Calculate to see its effect on Q. Repeat until you get Q very close to zero.

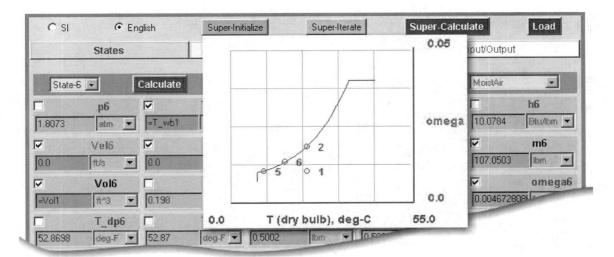

■ *HVAC Ex. #4*: **State-6 Evaluated**

A 12′ × 12′ × 10′ chamber at 1 atm ... With T6 assumed to be equal to T_wb1, State-6 is found without iteration. Plot the states on a psychrometric plot to show the three kinds of saturation.

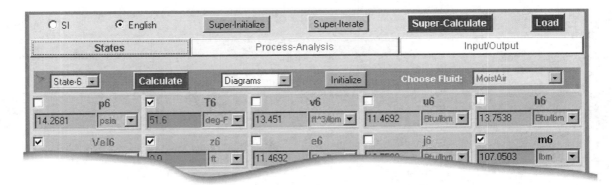

> ☐ **To converge to the adiabatic saturation temperature reduce T6 slightly below the T_wb1. Update Process-D. In this way, for T6=51.6 deg-F, Q calculated is only 4 Btu.**

■ *HVAC Ex. #4*: **State-6 Iterated**

A 12′ × 12′ × 10′ chamber at 1 atm ... After a few iterations (change T6, analyze Process-D), the refined value of T6 is only about 1 deg-F smaller than the wet bulb temperature of state-1.

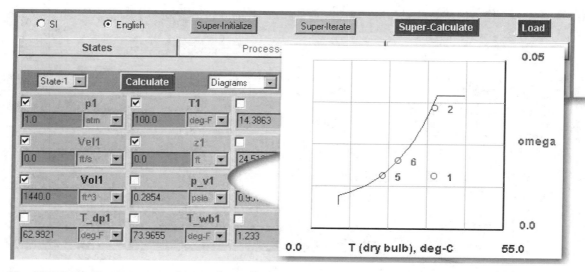

■ *HVAC Ex. #4:* **Parametric Study - Effect of T1**

A 12′ × 12′ × 10′ chamber at 1 atm ... In state-1, change the dry bulb temperature to 100 deg-F (and T6 to ′ = T_wb1′). Press the Enter key and then the Super-Calculate button to update all variables. The new psychrometric plot is displayed.

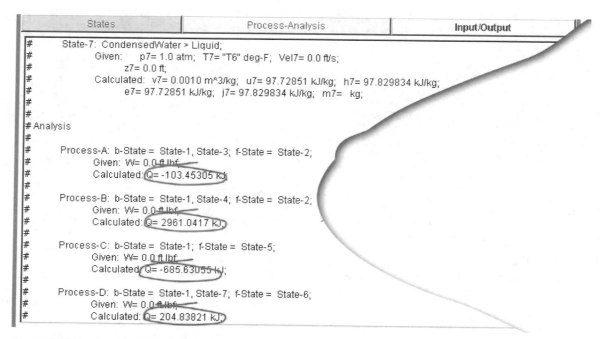

■ *HVAC Ex. #4:* **Detailed Solution Report**

A 12′ × 12′ × 10′ chamber at 1 atm ... The detailed report in the I/O panel generated by Super-Calculate operation contains the heat transfer for all the four processes updated for the new of T1.

```
####################################################################
# To regenerate this solution, copy the following TEST-Code onto the I/O panel of the
# ..Closed.Process.Specific.Cycles.IdealGas daemon
# and click the Load and Super-Calculate buttons.
#--------------------Start of TEST-Code-------------------------------------------------------
----

  States {
  State-1:  MoistAir;
  Given:       { p1= 1.0 atm;  T1= 70.0 deg-F;  Vel1= 0.0 ft/s;  z1= 0.0 ft;  Vol1= 1440.0 ft^3;
phi1= 30.0 %;  }

  State-2:  MoistAir;
  Given:       { T2= "T1" deg-F;  Vel2= 0.0 ft/s;  z2= 0.0 ft;  Vol2= "Vol1" ft^3;  phi2= 100.0 %;  }

  State-3:  SaturatedSteam;
  Given:       { p3= 1.0 atm;  Vel3= 0.0 ft/s;  z3= 0.0 ft;  }

  State-4:  CondensedWater;
  Given:       { p4= "p1" atm;  T4= "T1" deg-F;  Vel4= 0.0 ft/s;  z4= 0.0 ft;  }

  State-5:  MoistAir;
  Given:       { p5= "p1" atm;  Vel5= 0.0 ft/s;  z5= 0.0 ft;  Vol5= "Vol1" ft^3;  phi5= 100.0 %;  }

  State-6:  MoistAir;
  Given:       { T6= "T_wb1" deg-F;  Vel6= 0.0 ft/s;  z6= 0.0 ft;  Vol6="Vol1" ft^3;  phi6=
100.0 %  }

  State-7:  CondensedWater;
  Given:       { p7= 1.0 atm;  T7= "T6" deg-F;  Vel7= 0.0 ft/s;  z7= 0.0 ft;  }
  }

  Analysis {
  Process-A: b-State = State-1, State-3;  f-State = State-2;
  Given: { W= 0.0 ft.lbf;  }

  Process-B: b-State = State-1, State-4;  f-State = State-2;
  Given: { W= 0.0 ft.lbf;  }

  Process-C: b-State = State-1;  f-State = State-6;
  Given: { W= 0.0 ft.lbf;  }

  Process-D: b-State = State-1, State-7;  f-State = State-6;
  Given: { W= 0.0 ft.lbf;  }
  }
```

■ *HVAC Ex. #4:* **TEST-Code to Regenerate Visual Solution**

Combustion

Like the HVAC daemons, combustion may involve a closed process or a steady open device. Accordingly, the **Combustion** daemons can be found both under the closed and open system branches of the TEST-Map. Unlike other daemons, which are built around the state panel, a Reaction panel where a balanced reaction is created constitutes the core of the Combustion daemons. Once a reaction is established, the states of the fuel, oxidizer, and products are found separately. Calculated states are automatically loaded in the Analysis panel. Results from the analysis panel (for instance, j or e in the case of the evaluation of the adiabatic fram temperature), however, must be manually copied to the appropriate state panel.

Both the perfect gas and ideal gas models can be used for the gaseous species. The Reaction panel can be used not only to balance a theoretical reaction, but also for dry analysis, reaction with excess or deficient air, and reaction with a fuel mixture. Just as the SI and English unit buttons can be used to convert units, the Molar and Mass buttons toggle the basis of a reaction between mass and mol, allowing all four combinations (kg, kmol, lbm, and lbmol). A number of examples, besides the ones presented here, can be found in the Tutorial and in Chapter 14 of the Archive.

Example: Methane (CH$_4$) gas enters a steady-flow adiabatic combustor at 25 deg C and 1 atm. It is burned with 30% excess air, which also enters at the same temperature and pressure. Assuming complete combustion and no pressure drop, determine (a) the temperature of the products and (b) the entropy generation rate if the mass flow rate of the fuel is 1 kg/s.

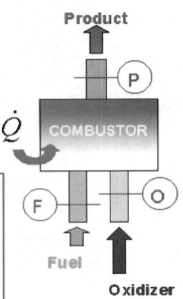

▣ Solution Procedure: Start the open, steady combustion daemon. Balance the reaction. Evaluate the fuel, oxidizer, and products states. They are auto-exported to the analysis panel. If a state variable is determined by the energy equation, manually copy and paste it back to the appropriate state, and evaluate the state completely.

■ *Combustion Ex. #1:* **Problem Statement**

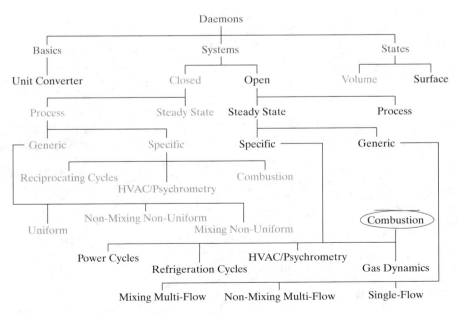

■ *Combustion Ex. #1:* **Simplify the Problem**

Methane (C4) gas enters... The appropriate daemon page is Daemons.Systems.Open.SteadyState. Specific.Combustion. Navigate systematically or use the Map.

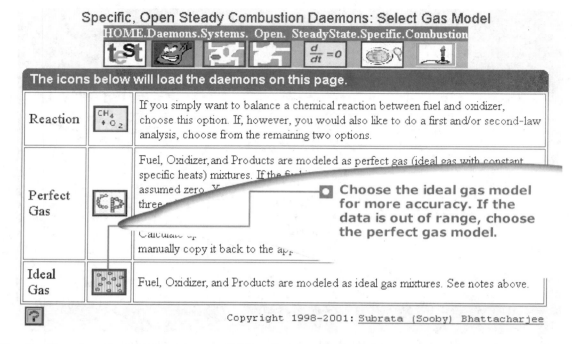

Copyright 1998-2001: Subrata (Sooby) Bhattacharjee

■ *Combustion Ex. #1*: **Select a Model for the Working Fluid**

Methane (CH4) gas enters... Choose the ideal gas model for better accuracy (variable c_p).

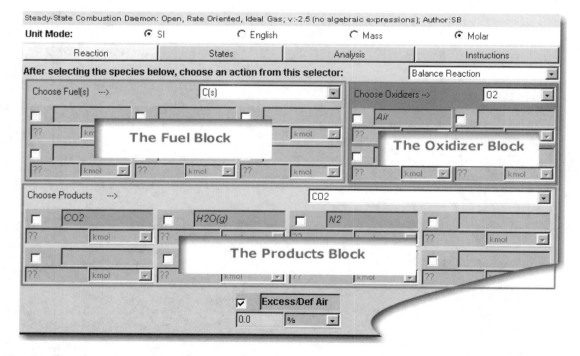

■ *Combustion Ex. #1*: **Reaction Panel**

Methane (CH4) gas enters... Select the appropriate radio buttons, SI, and Molar. The reaction panel has three blocks of species selectors, with the oxidizer and products blocks having some default choices.

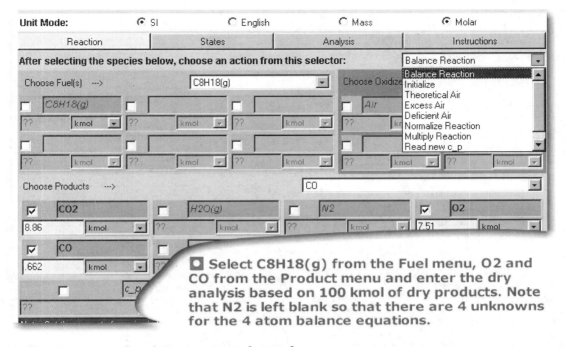

□ Select C8H18(g) from the Fuel menu, O2 and CO from the Product menu and enter the dry analysis based on 100 kmol of dry products. Note that N2 is left blank so that there are 4 unknowns for the 4 atom balance equations.

■ *Combustion Ex. #1:* **Choose CH4 as the Fuel**

Methane (CH4) gas enters... You may select more than one fuel if a fuel mixture is given.

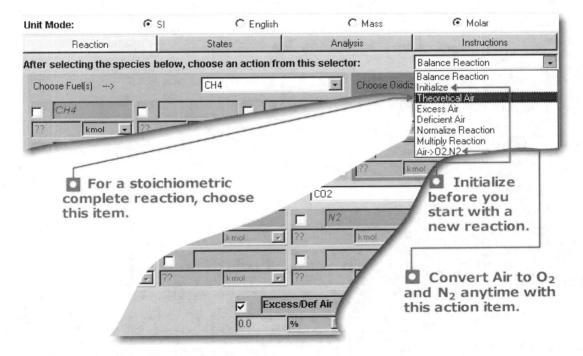

□ For a stoichiometric complete reaction, choose this item.

□ Initialize before you start with a new reaction.

□ Convert Air to O₂ and N₂ anytime with this action item.

■ *Combustion Ex. #1:* **The Action Menu**

Methane (CH4) gas enters... The pull-down action menu contains a number of action items. For a theoretical reaction, select the fuel (leave the oxidizer and products blocks alone) and select Theoretical Air to get the balanced reaction.

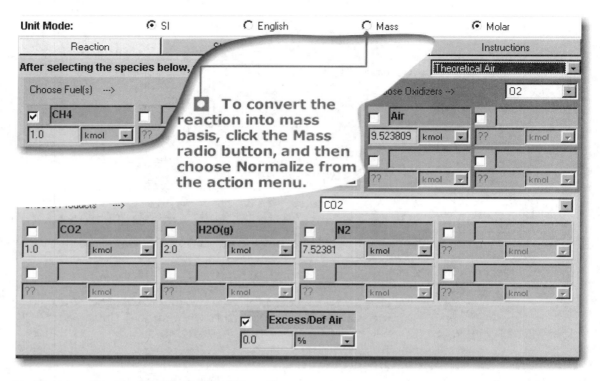

Combustion Ex. #1: **Theoretical Reaction**

Methane (CH4) gas enters... The theoretical reaction is expressed on the molal basis.

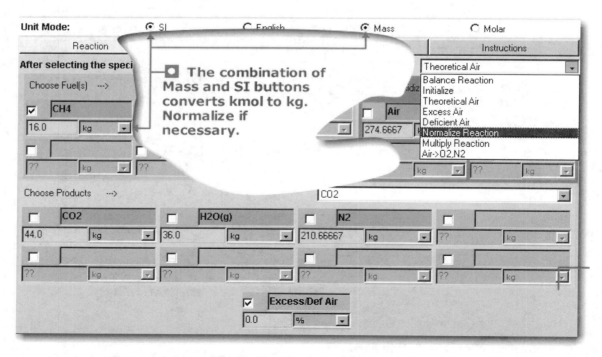

Combustion Ex. #1: **Converting from Mole to Mass Basis**

Methane (CH4) gas enters... Use the combination SI (or English) and Mass to express the reaction in terms of mass (kg or lbm). Normalize Reaction to convert to unit mass.

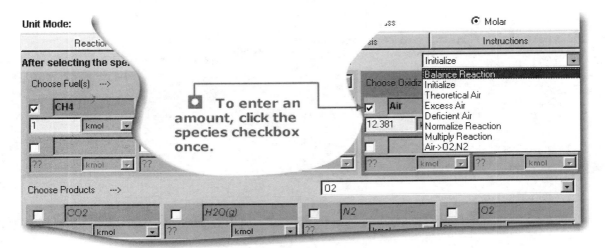

☑ **For 30% excess air, enter 1.3*9.52=12.38 kmol of Air, enter 1 kmol of CH₄, select O₂ in the product block, and choose Balance Reaction from the action menu.**

■ *Combustion Ex. #1:* **Balancing the Reaction for 30% Excess All**

Methane (CH4) gas enters... The stoichiometric coefficient of air is multiplied by 1.3 to obtain the amount of air in the fuel lean reaction.

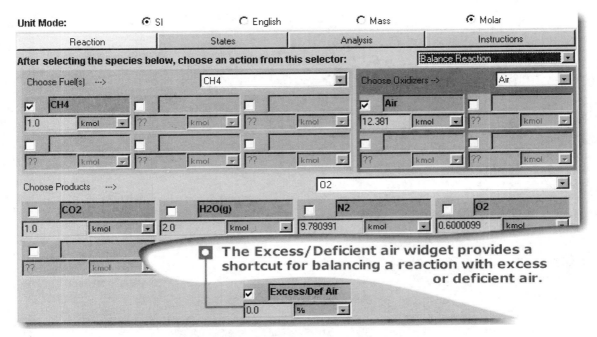

■ *Combustion Ex. #1:* **Reaction Balanced on Mole Basis**

Methane (CH4) gas enters... Balancing a reaction uses an atomic balance. With four unique atoms in this reaction, the coefficient of any four of the six species (reactants and products) can be determined.

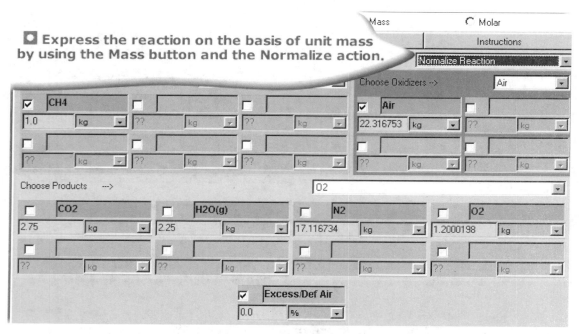

Express the reaction on the basis of unit mass by using the Mass button and the Normalize action.

■ *Combustion Ex. #1:* **Reaction Converted to Mass Basis**

Methane (CH4) gas enters... The normalized reaction is expressed in terms of unit mass by selecting SI and Mass buttons and then applying the Normalize Reaction action item.

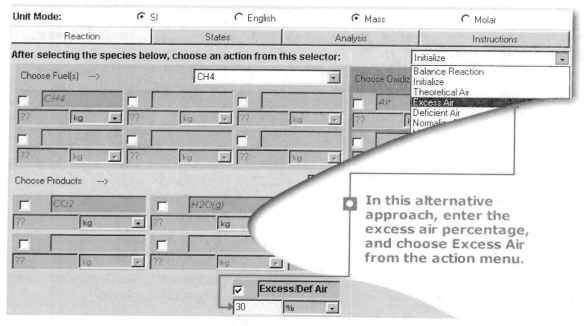

In this alternative approach, enter the excess air percentage, and choose Excess Air from the action menu.

■ *Combustion Ex. #1:* **Balancing with the Help of the Input Widget**

Methane (CH4) gas enters... Excess or deficient air percentage can be directly entered and the reaction balanced using the appropriate action item.

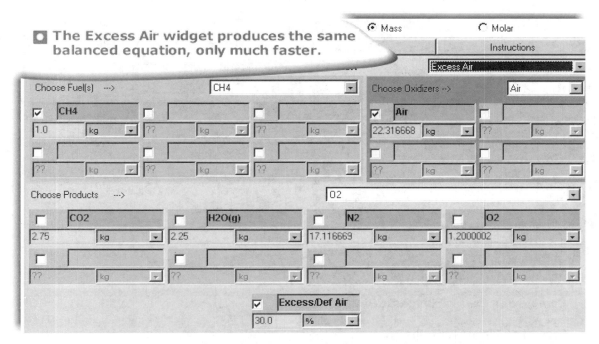

The Excess Air widget produces the same balanced equation, only much faster.

■ *Combustion Ex. #1:* **Use of Excess Air Widget Results in the Same Reaction**

Methane (CH4) gas enters... The balanced equation is based on 30% excess air.

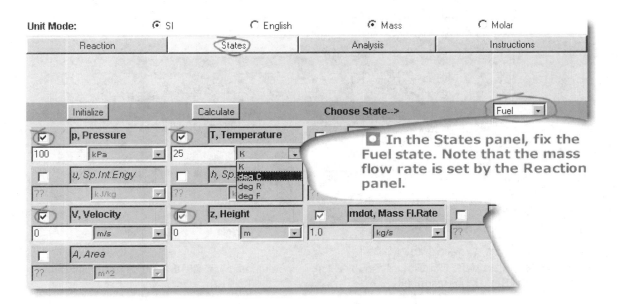

In the States panel, fix the Fuel state. Note that the mass flow rate is set by the Reaction panel.

■ *Combustion Ex. #1:* **Evaluate the Fuel State**

Methane (CH4) gas enters... Switch to the States panel, and work on the fuel, oxidizer, and products states.

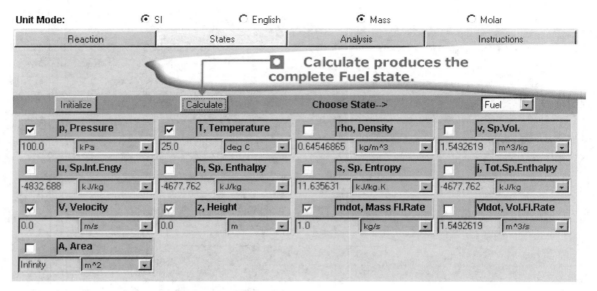

■ *Combustion Ex. #1:* **Fuel State Found**

Methane (CH4) gas enters... The fuel state is completely evaluated from the pressure and temperature. The mass flow rate is imported from the Reaction panel.

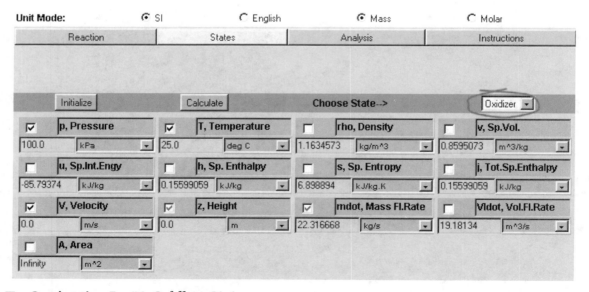

■ *Combustion Ex. #1:* **Oxidizer State**

Methane (CH4) gas enters... Evaluate the oxidizer state in a similar manner. The flow area is infinity because the velocity of the flow (ke) is ignored.

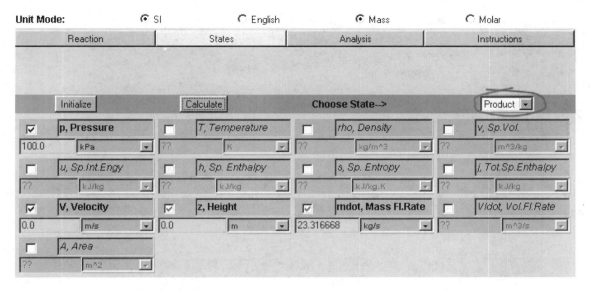

Combustion Ex. #1: **Products State**

Methane (CH4) gas enters. . . With the temperature being the desired unknown, the products state is only partially evaluated.

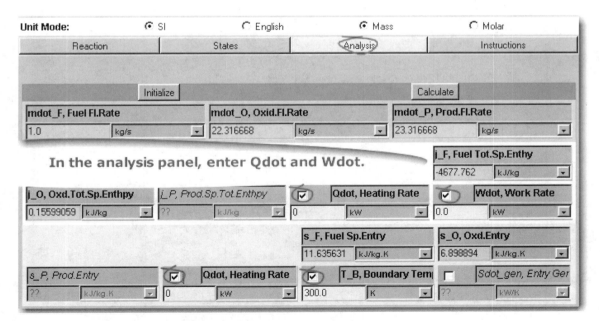

Combustion Ex. #1: **The Analysis Panel**

Methane (CH4) gas enters. . . The mass equation is already satisfied due to a balanced reaction. Enter Qdot and Wdot in the energy panel.

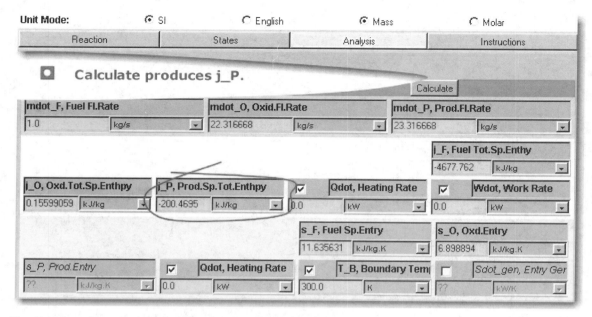

Combustion Ex. #1: **The Product Enthalpy**

Methane (CH4) gas enters... Copy j_P into the product state manually. Note that unlike other daemons, the combustion daemons do not automatically post the results in the state panels.

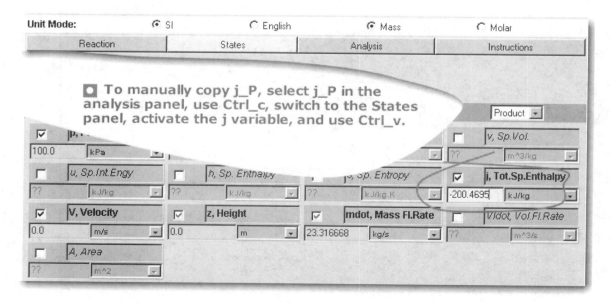

Combustion Ex. #1: **Copy and Paste j_p onto the Products Panel**

Methane (CH4) gas enters... The combustion daemons currently do not support automatic posting of the state variables. They also do not support equations.

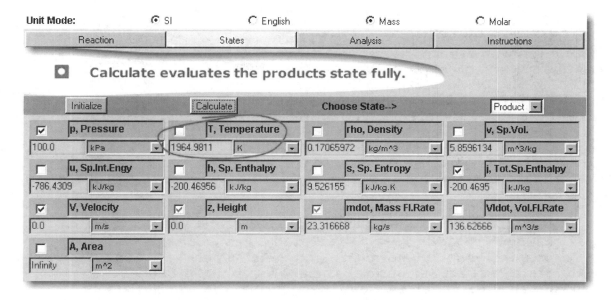

Combustion Ex. #1: **Products State Evaluated**

Methane (CH4) gas enters... The adiabatic flame temperature is now found as part of the products state.

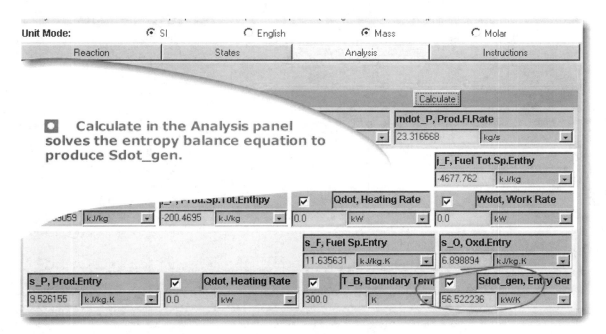

Combustion Ex. #1: **Entropy Generation Rate Found**

Methane (CH4) gas enters... Now that the products state is fully known, the entropy balance equation can be solved for Sdot_gen in the Analysis panel using the Calculate button again.

Example: Octane (C_8H_{18}) in gaseous form is burned with dry air. The volumetric analysis of the products on a dry basis is 8.86% CO_2, 0.662% CO, 7.51% O_2, and 82.978% N_2. (a) Determine the air–fuel ratio. (b) If the initial pressure and temperature of the air–fuel mixture are 100 kPa and 25 deg C, determine the final pressure. Assume the combustion chamber to be an insulated, closed tank containing 2 kg of fuel at the start.

> ◘ **Solution Procedure: Start the Closed Combustion daemon. Balance the equation, normalize, and then multiply by 2. Evaluate the states as best as possible. The analysis panel produces unknown state variables. Copy them back to the appropriate state, and calculate the products state completely.**

■ *Combustion Ex. #2:* **Problem Description and Solution Algorithm**

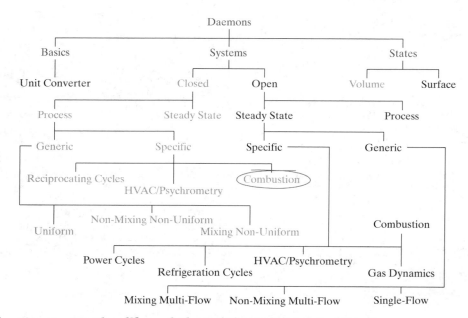

■ *Combustion Ex. #2:* **Simplify and Choose the Appropriate Daemon**

Octane (C8H18) in gaseous form is burned with. . . The appropriate daemon is Daemons.Systems.Closed.Process.Specific.Combustion.

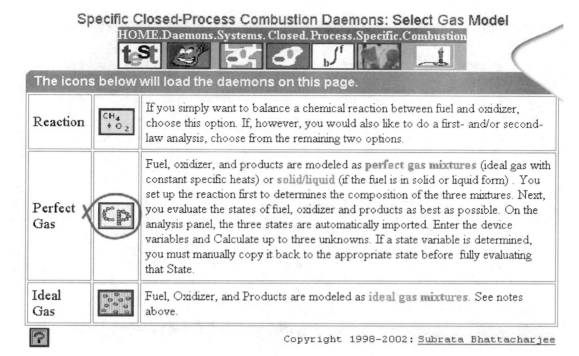

Copyright 1998–2002: Subrata Bhattacharjee

■ *Combustion Ex. #2:* **Select a Working Fluid Model**

Octane (C8H18) in gaseous form is burned with... We choose the perfect gas model because of its simplicity in a manual solution.

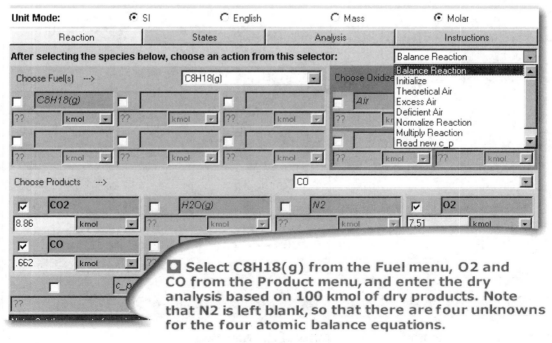

☐ Select C8H18(g) from the Fuel menu, O2 and CO from the Product menu, and enter the dry analysis based on 100 kmol of dry products. Note that N2 is left blank, so that there are four unknowns for the four atomic balance equations.

■ *Combustion Ex. #2:* **Enter the Dry Analysis Results**

Octane (C8H18) in gaseous form is burned with... With four atomic balance equations, four coefficients can be solved.

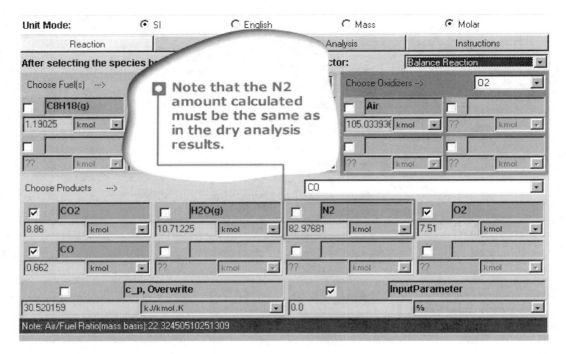

■ Combustion Ex. #2: **Balance Reaction**

Octane (C8H18) in gaseous form is burned with... The balanced reaction is expressed in terms of 1.19 kmol of fuel. To convert the basis to 2 kg of fuel, normalize and multiply the reaction suitably.

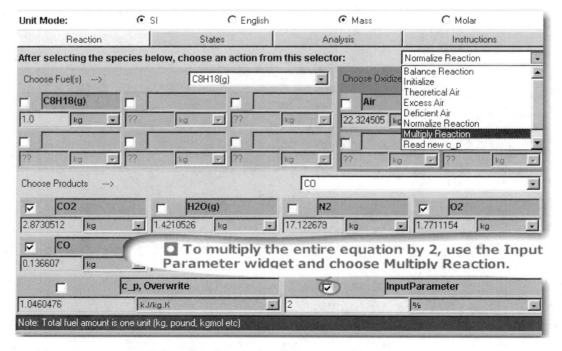

■ Combustion Ex. #2: **Make 2 kg the Basis of the Reaction**

Octane (C8H18) in gaseous form is burned with... The final reaction must be based on 2 kg of fuel. Select the SI and Mass buttons, Normalize the reaction, and then multiply by 2 using the Input Parameter variable.

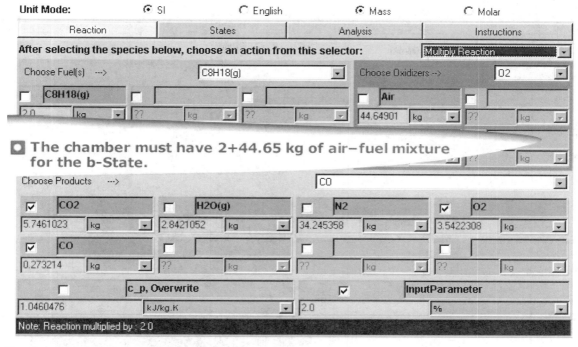

Combustion Ex. #2: **The Balanced Reaction**

Octane (C8H18) in gaseous form is burned with... Now that the reaction is balanced, we move on to the States panel and try to evaluate the Fuel, Oxidizer, and Products states.

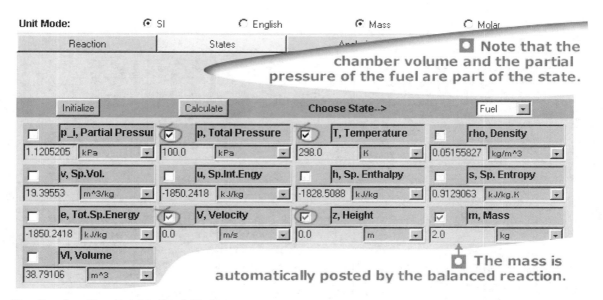

Combustion Ex. #2: **Fuel State**

Octane (C8H18) in gaseous form is burned with... The fuel state can be completely evaluated from the total pressure and temperature.

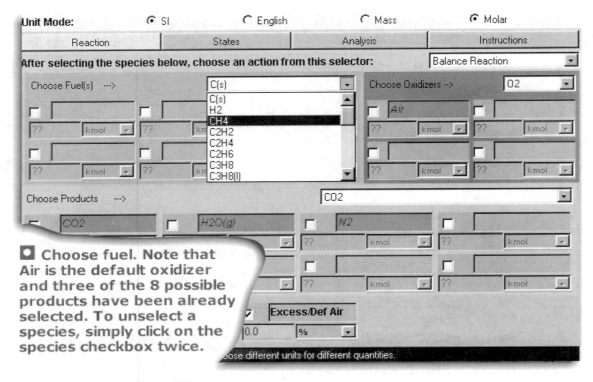

■ Choose fuel. Note that Air is the default oxidizer and three of the 8 possible products have been already selected. To unselect a species, simply click on the species checkbox twice.

■ Combustion Ex. #2: Oxidizer State

Octane (C8H18) in gaseous form is burned with... The oxidizer state, too, can be completely evaluated from the total pressure and temperature.

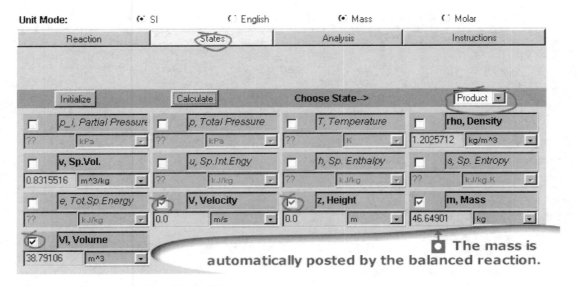

■ The mass is automatically posted by the balanced reaction.

■ Combustion Ex. #2: Products State

Octane (C8H18) in gaseous form is burned with... Manually copy the chamber volume to the products state, and partially Calculate the state.

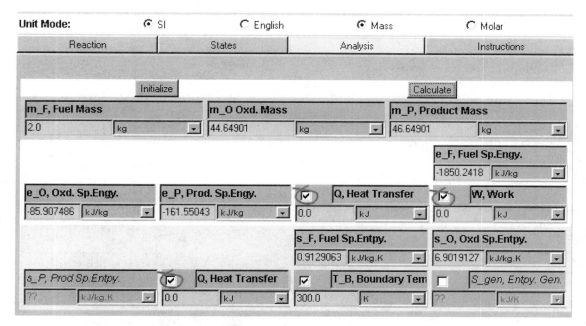

■ *Combustion Ex. #2*: **Analysis Panel**

Octane (C8H18) in gaseous form is burned with. . . Enter Q = 0 and W = 0. Calculate produces e_P. Manually copy and paste it back to the Products state.

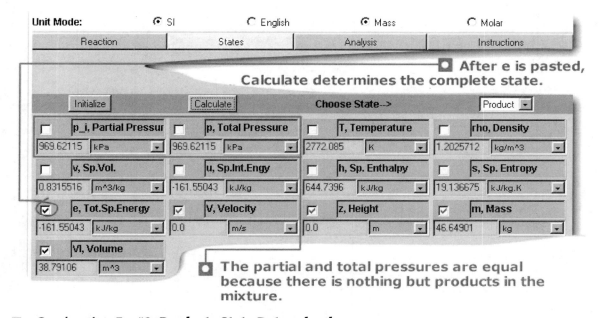

■ *Combustion Ex. #2*: **Products State Determined**

Octane (C8H18) in gaseous form is burned with. . . With e_P, the state is completely evaluated. The final pressure and temperature are found.

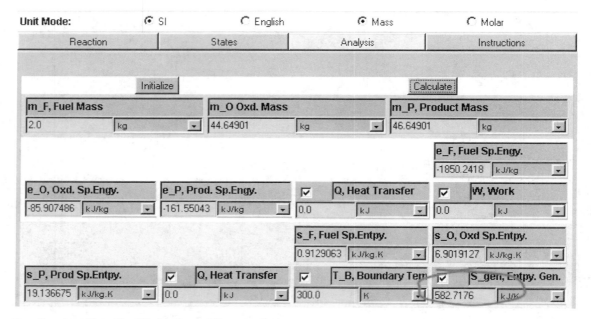

■ *Combustion Ex. #2:* **Entropy Generation**

Octane (C8H18) in gaseous form is burned with... In the Analysis panel, Calculate now produces entropy generation during the process.

Gas Dynamics

The **Gas Dynamics** daemon is an extension of the Open Steady daemons with the perfect gas model for the working fluid. The high-speed flows introduce additional state variables, such as the stagnation properties and the Mach number, which are reflected on the modified State panel. The Analysis panel remains just the same as the Open Steady daemon. An additional layer under the heading **Tables** that contains the isentropic flow table, the normal shock table, the delta–theta oblique-shock charts, and the Prandtl–Meyer functions—the last two tables used mostly at the graduate level. For problems involving a working fluid other than gases (say, steam), the Open Steady daemons can be used. A number of additional examples can be found in the Tutorial and in Chapter 15 of the Archive.

Example: Air at 1 MPa and 550 deg C enters a converging nozzle with a velocity of 100 m/s. Determine the exit Mach number for a back pressure of (a) 0.7 MPa, (b) 0.4 MPa, and (c) 0.2 MPa.

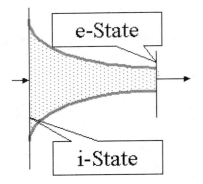

> 🔲 **Solution Procedure: Start the gas dynamics daemon. Evaluate the inlet state. Three different ways of determining the exit state are illustrated in this example.**

■ *Gas Dynamics Ex. #1:* **Problem Description**

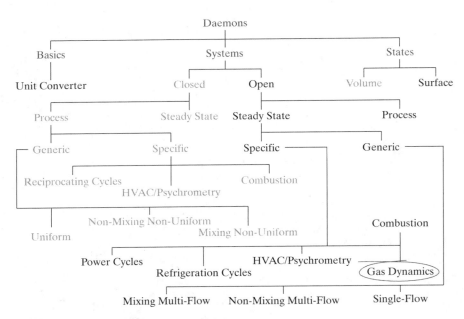

■ *Gas Dynamics Ex. #1:* **Simplify the Problem**

Air at 1 MPa and 550 deg-C enters a nozzle...The appropriate daemon is... Systems.Open. SteadyState.Specific.GasDynamics. Navigate systematically or use the Map.

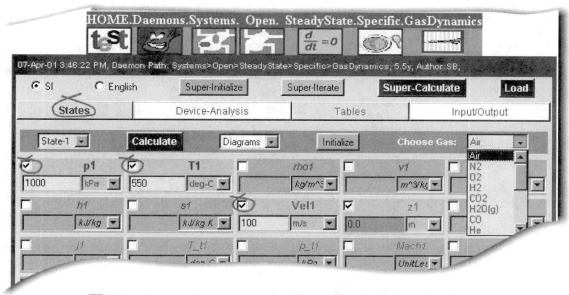

☐ **Perfect gas is the working fluid model. Choose the gas and enter the known variables.**

■ *Gas Dynamics Ex. #1*: Evaluate the Inlet State (State-1)

Air at 1 MPa and 550 deg-C enters a nozzle... Air is the default gas. Choose State-1 to represent the inlet state.

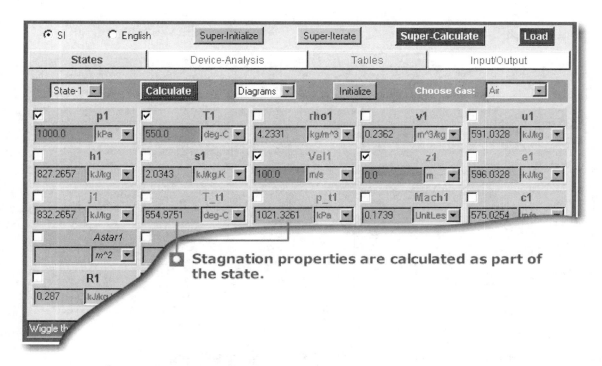

☐ **Stagnation properties are calculated as part of the state.**

■ *Gas Dynamics Ex. #1*: Inlet State Evaluated

Air at 1 MPa and 550 deg-C enters a nozzle... Calculate the state. Variable Astar cannot be calculated because the mass flow rate is not known.

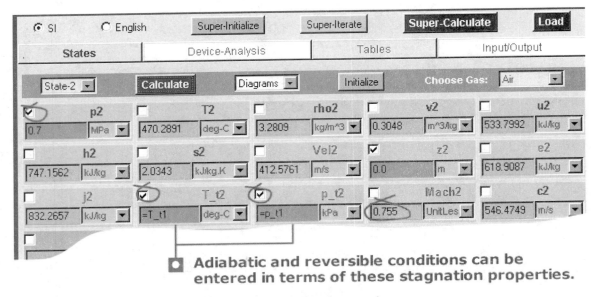

Adiabatic and reversible conditions can be entered in terms of these stagnation properties.

■ *Gas Dynamics Ex. #1*: Evaluate the Exit State (State-2)

Air at 1 MPa and 550 deg-C enters a nozzle... Enter the back pressure as the pressure at the exit (p2), and enter the stagnation properties. Calculate produces M = .755, the subsonic speed validating the use of p2=p_back.

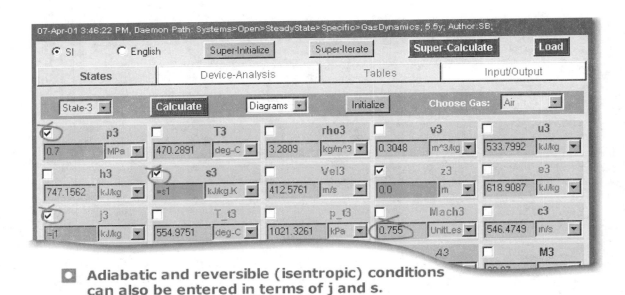

Adiabatic and reversible (isentropic) conditions can also be entered in terms of j and s.

■ *Gas Dynamics Ex. #1*: Evaluate the Exit State in a Different Way

Air at 1 MPa and 550 deg-C enters a nozzle... Instead of stagnation properties, the traditional variables, j and s, could be used to satisfy energy and entropy balance equations. State-3 can be seen to be identical to State-2 (M=0.755).

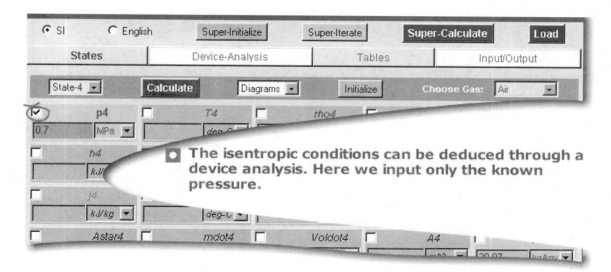

The isentropic conditions can be deduced through a device analysis. Here we input only the known pressure.

■ *Gas Dynamics Ex. #1*: Evaluate the Exit State with a Third Approach

Air at 1 MPa and 550 deg-C enters a nozzle. . . Partially evaluate State-4, with only pressure known at this point. We will do a steady flow analysis to deduce j4 = j1 and s4 = s1 .

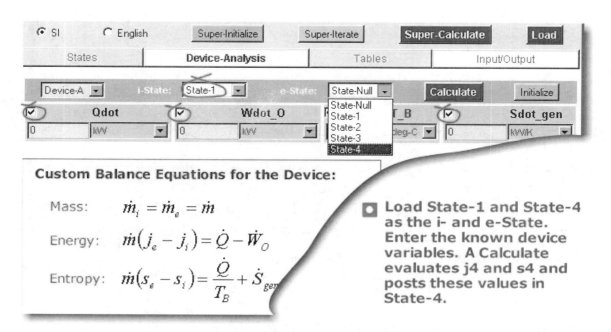

Custom Balance Equations for the Device:

Mass: $\dot{m}_i = \dot{m}_e = \dot{m}$

Energy: $\dot{m}(j_e - j_i) = \dot{Q} - \dot{W}_O$

Entropy: $\dot{m}(s_e - s_i) = \dfrac{\dot{Q}}{T_B} + \dot{S}_{gen}$

Load State-1 and State-4 as the i- and e-State. Enter the known device variables. A Calculate evaluates j4 and s4 and posts these values in State-4.

■ *Gas Dynamics Ex. #1*: Analyze the Steady Single-Flow Device

Air at 1 MPa and 550 deg-C enters a nozzle. . . Load State-1 and State-4 as the inlet and exit states, and enter Qdot (adiabatic) and Wdot_O. Calculate. Variables calculated, j_e and s_e, are posted back to State-4.

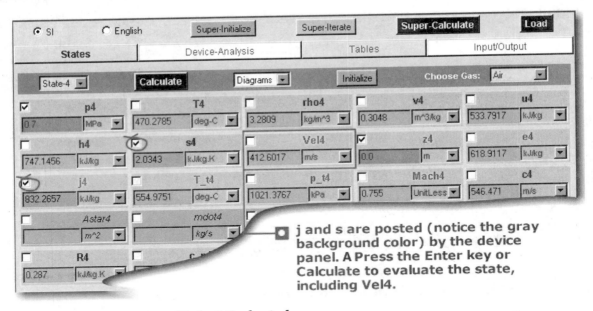

j and s are posted (notice the gray background color) by the device panel. A Press the Enter key or Calculate to evaluate the state, including Vel4.

■ Gas Dynamics Ex. #1: State-4 Evaluated

Air at 1 MPa and 550 deg-C enters a nozzle. . . Variables j, s, and mdot are posted back to State-4. Calculate produces M=0.755, identical to that obtained by the direct methods. Note that the last two approaches are more general and can handle any fluid model, rather than the simple perfect gas model applied in gas dynamic formulas.

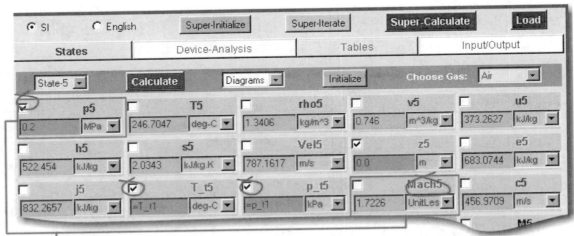

An exit pressure of 0.2 MPa produces a supersonic Mach number, which is impossible for an isentropic converging nozzle. The actual exit Mach number, therefore, must be one.

■ Gas Dynamics Ex. #1: Exit Mach Number at 0.2 MPa

Air at 1 MPa and 550 deg-C enters a nozzle. . . For an exit pressure of 0.2 MPa, the exit velocity is supersonic, an impossibility with a converging nozzle. Therefore, the exit Mach number must be 1, the maximum permissible value for a converging nozzle.

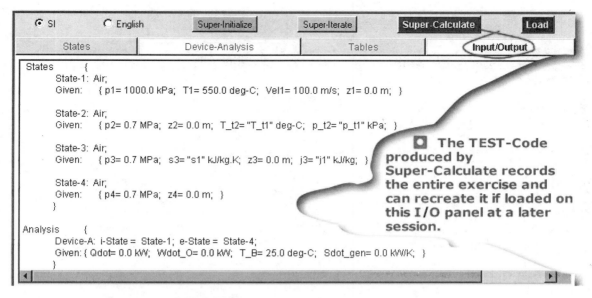

```
States      {
     State-1: Air;
     Given:    { p1= 1000.0 kPa;  T1= 550.0 deg-C;  Vel1= 100.0 m/s;  z1= 0.0 m;  }

     State-2: Air;
     Given:    { p2= 0.7 MPa;  z2= 0.0 m;  T_t2= "T_t1" deg-C;  p_t2= "p_t1" kPa;  }

     State-3: Air;
     Given:    { p3= 0.7 MPa;  s3= "s1" kJ/kg.K;  z3= 0.0 m;  j3= "j1" kJ/kg;  }

     State-4: Air;
     Given:    { p4= 0.7 MPa;  z4= 0.0 m;  }
     }

Analysis    {
     Device-A: i-State = State-1;  e-State = State-4;
     Given: { Qdot= 0.0 kW;  Wdot_O= 0.0 kW;  T_B= 25.0 deg-C;  Sdot_gen= 0.0 kW/K;  }
     }
```

🔲 **The TEST-Code produced by Super-Calculate records the entire exercise and can recreate it if loaded on this I/O panel at a later session.**

■ *Gas Dynamics Ex. #1*: TEST-Codes

Air at 1 MPa and 550 deg-C enters a nozzle. . . . The Super-Calculate operation generates the TEST-Codes and a solution report on this I/O panel.

Example: Determine the velocity of sound in steam at 1000 kPa and 400 deg C. What-if Scenario: How does the speed change if the pressure is 100 kPa instead?

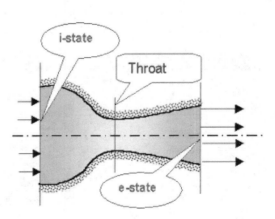

🔲 **Solution Procedure: The velocity of sound in a medium is given by the square root of the partial derivative of p with respect to rho of the medium while entropy is held constant. We will evaluate two neighboring states around the given conditions and obtain the partial derivative by using its definition.**

■ *Gas Dynamics Ex. #2*: Problem Statement

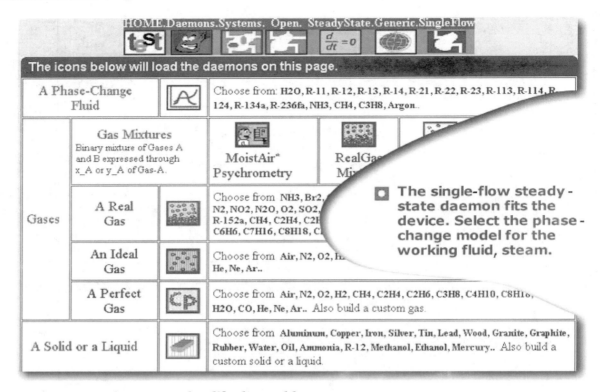

The single-flow steady-state daemon fits the device. Select the phase-change model for the working fluid, steam.

Gas Dynamics Ex. #2: Simplify the Problem

Determine the velocity of sound in steam. . . The gas dynamics daemon is based on the perfect gas model. So we proceed systematically, and the simplification leads to the following generic daemon page: . . . Systems. Open.SteadyState.Generic.Single Flow. Select the phase-change model for steam.

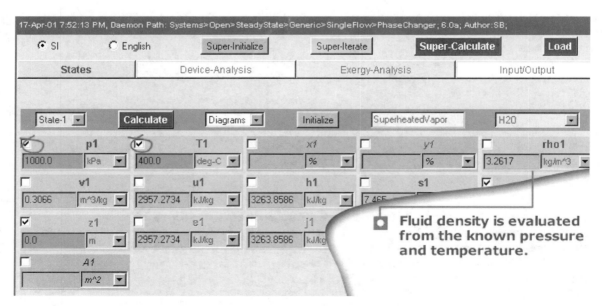

Fluid density is evaluated from the known pressure and temperature.

Gas Dynamics Ex. #2: Evaluate the State at 1000 kPa. 400 deg. . . C (State 1)

Determine the velocity of sound in steam. . . The state panel of the generic daemons does not contain gas dynamic variables such as the Mach number. However, we can evaluate it from the first principle.

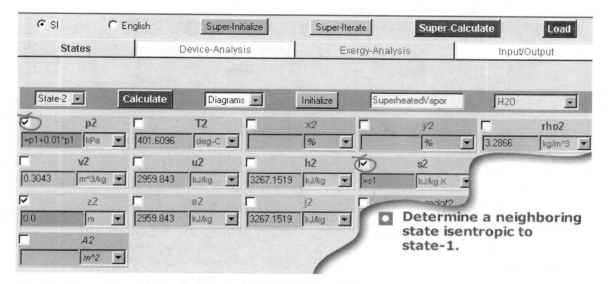

Determine a neighboring state isentropic to state-1.

■ *Gas Dynamics Ex. #2:* Evaluate a Neighboring State

Determine the velocity of sound in steam... State-2 is isentropic to State-1 (s2=s1) and is in its close neighborhood (p2 is only 1% higher than p1).

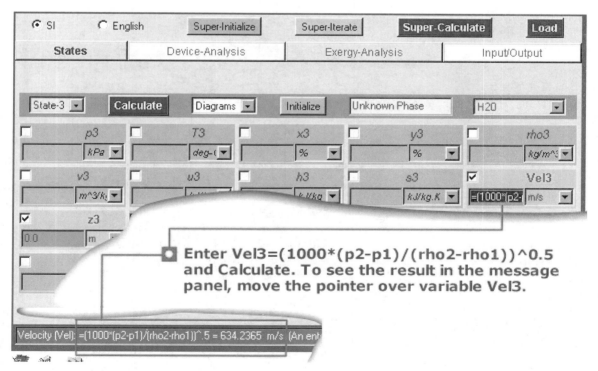

Enter Vel3=(1000*(p2-p1)/(rho2-rho1))^0.5 and Calculate. To see the result in the message panel, move the pointer over variable Vel3.

Velocity (Vel): =(1000*(p2-p1)/(rho2-rho1))^.5 = 634.2365 m/s (An ent

■ *Gas Dynamics Ex. #2:* The Sonic Velocity in Steam

Determine the velocity of sound in steam... We use State-3 simply to evaluate the speed of sound from the square root of the partial derivative of p with respect to rho. As you type in the algebraic expression, it also appears on the Message Panel, making it easier to edit.

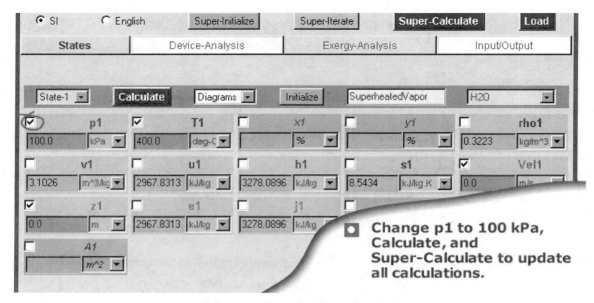

Change p1 to 100 kPa, Calculate, and Super-Calculate to update all calculations.

■ *Gas Dynamics Ex. #2:* Parametric Study—Reduce Pressure

Determine the velocity of sound in steam... To see the effect of pressure on the speed of sound, change the pressure to 100 kPa from 1000 kPa in State-1. Press the Enter key to register the change and the Super-Calculate button to update all panels.

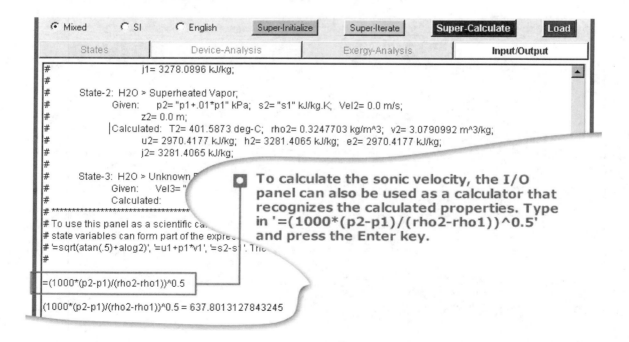

To calculate the sonic velocity, the I/O panel can also be used as a calculator that recognizes the calculated properties. Type in '=(1000*(p2-p1)/(rho2-rho1))^0.5' and press the Enter key.

■ *Gas Dynamics Ex. #2:* The New Sonic Velocity

Determine the velocity of sound in steam... Switch to State-3 and hover the pointer over Vel3. The new velocity, 637 m/s, appears in the Message Panel.

```
#####################################################################
# To regenerate this solution, copy the following TEST-Code onto the I/O panel of the
# ..Systems.Open.SteadyState.Generic.SingleFlow.Phase-Change daemon
# and click the Load and Super-Calculate buttons.
#--------------------Start of TEST-Code------------------------------------------------------
----

States {
State-1:  H2O;
Given:        { p1= 1000.0 kPa;  T1= 400.0 deg-C;  Vel1= 0.0 m/s;  z1= 0.0 m;  }

State-2:  H2O;
Given:        { p2= "p1+.01*p1" kPa;  s2= "s1" kJ/kg.K;  Vel2= 0.0 m/s;  z2= 0.0 m;  }

State-3:  H2O;
Given:        { Vel3= "(1000*(p2−p1)/(rho2−rho1))^.5" m/s;  z3= 0.0 m;  }
}
```

■ *Gas Dynamics Ex. #2:TEST-Code to Generate the Visual Solution*

Example: A rocket nozzle has a ratio of exit to throat area of 4.0. Stagnation temperature and pressure in the chamber are 1500 kPa and 3 MPa. Assume the gas to behave as a perfect gas with k=1.35 and molecular mass =20. Determine the exhaust velocity for isentropic nozzle flow and for the case when a normal shock is located just inside the exit plane.

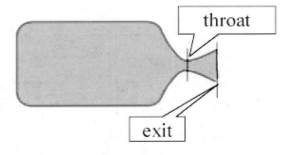

■ **Solution Procedure: Start the Gas Dynamics daemon. Choose a 'Custom' gas. Molecular mass and k are treated as user-defined material properties. The use of the built-in normal shock table is illustrated in this example.**

■ *Gas Dynamics Ex. #3: Problem Statement*

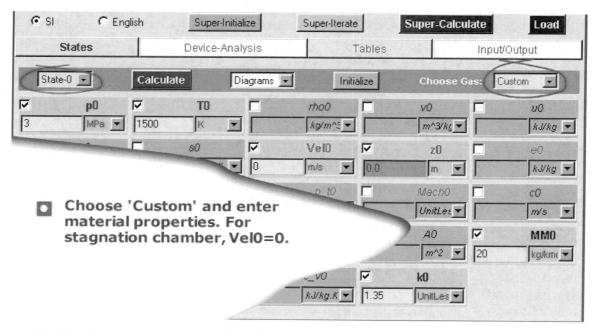

Gas Dynamics Ex. #3: Stagnation Chamber State (State-0)

A rocket nozzle has a ratio of exit to throat area of 4. . . Enter the custom properties of material, chamber pressure, the temperature, and velocity.

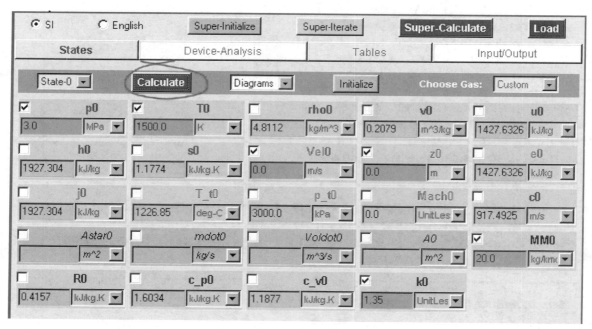

Gas Dynamics Ex. #3: Calculate the Stagnation State

A rocket nozzle has a ratio of exit to throat area of 4. . . Calculate produces the stagnation properties.

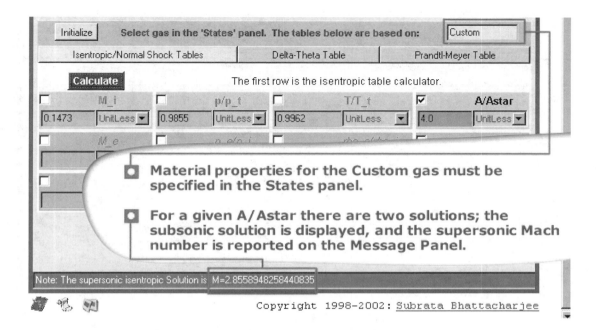

Gas Dynamics Ex. #3: Isentropic Table

A rocket nozzle has a ratio of exit to throat area of 4. . . Get to the Tables panel. The isentropic/normal shock table is the default table. Enter A/Astar and Calculate the Mach number. The supersonic solution appearing in the message panel is the correct solution for this converging-diverging nozzle.

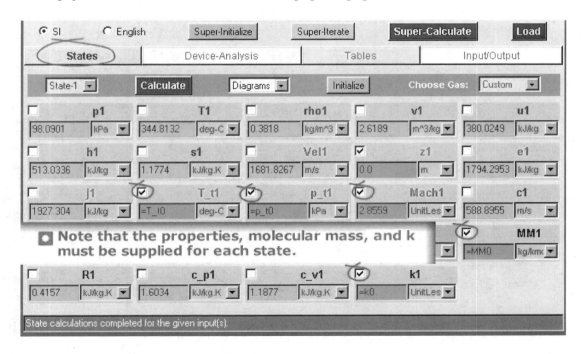

Gas Dynamics Ex. #3: Isentropic Exit State (State-1)

Ex. #3:. . .Determine the exhaust velocity for isentropic. . . Enter the stagnation temperature and pressure, and the Mach number. Because it is a custom gas, the material properties must be entered again. Calculate produces the isentropic exit velocity.

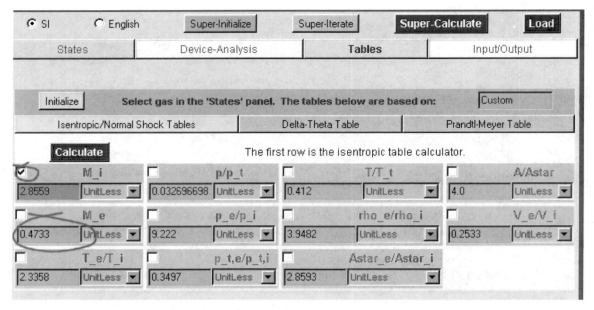

Gas Dynamics Ex. #3: Normal Shock Table

For a case when a normal shock is located at the exit. Enter the Mach number before the shock (the nozzle exit Mach number, M_i), and Calculate M_e, the Mach number after the shock. Also, the stagnation pressure ratio, p_t,e/p_t,i, is now known.

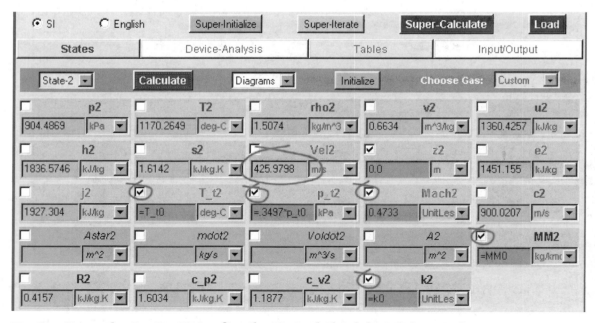

Gas Dynamics Ex. #3: State after the Normal Shock at Exit (State-2)

For a case when a normal shock is located at the exit. . . Use the Mach number and the stagnation pressure ratio calculated in the shock table. The total temperature remains unchanged through the shock, of while the stagnation pressure drops. Calculate produces a desired velocity of 426 m/s.

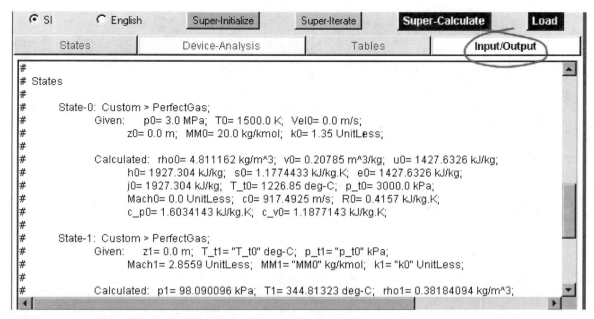

Gas Dynamics Ex. #3: Detailed Report

A rocket nozzle has a ratio of exit to throat area of 4... A Super-Calculate produces the detailed report and the TEST-Codes on the I/O panel.

```
###################################################################
# To regenerate this solution, copy the following TEST-Code onto the I/O panel of the
# ..Open.Steady.Specific.GasDynamics daemon
# and click the Load and Super-Calculate buttons.
#--------------------Start of TEST-Code---------------------------------------------------------
----

  States {
  State-0:  Custom;
  Given:        { p0= 3.0 MPa;  T0= 1500.0 K;  Vel0= 0.0 m/s;  z0= 0.0 m;  MM0= 20.0
kg/kmol;  k0= 1.35 UnitLess;  }

  State-1:  Custom;
  Given:        { z1= 0.0 m;  T_t1= "T_t1" deg-C;  p_t1= "p_t1" kPa;  Mach1= 2.8559
Unitless;  MM1= "MM1" kg/kmol;  k1= "k1"  UnitLess;  }

  State-2:  Custom;
  Given:        { z2= 0.0 m;  T_t2= "T_t1" deg-C;  p_t2= ".3497*p_t1" kPa;  Mach2= 0.4733
UnitLess;  MM2= "MM0" kg/kmol;  k2= "k1" UnitLess;  }
  }
```

Gas Dynamics Ex. #3: TEST-Code to Regenerate the Visual Solution

Other Daemons

There are quite a few daemons available in TEST that have not been discussed in this book. The Unit Converter is pretty much self-explanatory and has the unique feature of collecting user input to build up the database. There is a Black-body Radiation daemon that calculates the Planck function, and a Band Radiation daemon that calculates emissions at different bands of CO_2, H_2O and CO. Currently, there are no daemons for chem

INDEX

LICENSE AGREEMENT AND LIMITED WARRANTY

READ THIS LICENSE CAREFULLY BEFORE OPENING THE CD SOFTWARE PACKAGE. BY OPENING THE CD SOFTWARE PACKAGE, YOU ARE AGREEING TO THE TERMS AND CONDITIONS OF THIS LICENSE. IF YOU DO NOT AGREE, DO NOT OPEN THE CD SOFTWARE PACKAGE. *THESE TERMS APPLY TO ALL LICENSED SOFTWARE ON THE DISC EXCEPT THAT THE TERMS FOR USE OF ANY SHARE-WARE OR FREEWARE ON THE CD ARE AS SET FORTH IN THE RELEVANT LICENSE LOCATED IN THIS APPENDIX AND/OR IN THE RELEVANT ELECTRONIC LICENSE LOCATED ON THE CD ("FREEWARE LICENSES"):*

1. **GRANT OF LICENSE and OWNERSHIP:** The enclosed computer programs ("Software") are licensed, not sold, to you by Prentice-Hall, Inc. ("We" or the "Company") in consideration of your purchase of this book, and your agreement to these terms. You own only the disc but we and/or our licensors own the Software itself. This license grants to you a nonexclusive license to use and display the enclosed copy of the Software on a single computer (i.e. with a single CPU), for educational use only, so long as you comply with the terms of this Agreement. You may make one copy for back up only. We reserve any rights not granted to you. The applicable Freeware License may grant you additional rights as between you and the shareware authors.

2. **USE RESTRICTIONS:** You may <u>not</u> sell or license copies of the Software or the Documentation to others, except as provided in any applicable Freeware License. You may <u>not</u> transfer, distribute or make available the Software or the Documentation, except as provided in any applicable Freeware License. You may <u>not</u> reverse engineer, disassemble, decompile, modify, adapt, translate or create derivative works based on the Software or the Documentation, except as provided in any applicable Freeware License. You may be held legally responsible for any copying or copyright infringement which is caused by your failure to abide by the terms of these restrictions.

3. **TERMINATION:** This license is effective until terminated. This license will terminate automatically without notice from the Company if you fail to comply with any provisions or limitations of this license. Upon termination, you shall destroy the Documentation and all copies of the Software. All provisions of this Agreement as to limitation and disclaimer of warranties, limitation of liability, remedies or damages, and our ownership rights shall survive termination.

4. **DISCLAIMER OF WARRANTY: THE COMPANY AND ITS LICENSORS MAKE NO WARRANTIES ABOUT THE SOFTWARE, WHICH IS PROVIDED "AS-IS." IF THE DISC IS DEFECTIVE IN MATERIALS OR WORKMANSHIP, YOUR ONLY REMEDY IS TO RETURN IT TO THE COMPANY WITHIN 30 DAYS FOR REPLACEMENT UNLESS THE COMPANY DETERMINES IN GOOD FAITH THAT THE DISC HAS BEEN MISUSED OR IMPROPERLY INSTALLED, REPAIRED, ALTERED OR DAMAGED. THE COMPANY DISCLAIMS ALL WARRANTIES, EXPRESS OR IMPLIED, INCLUDING WITHOUT LIMITATION, THE IMPLIED WARRANTIES OF MERCHANTABILITY AND FITNESS FOR A PARTICULAR PURPOSE. THE COMPANY DOES NOT WARRANT, GUARANTEE OR MAKE ANY REPRESENTATION REGARDING THE ACCURACY, RELIABILITY, CURRENTNESS, USE, OR RESULTS OF USE, OF THE SOFTWARE.**

5. **LIMITATION OF REMEDIES AND DAMAGES: IN NO EVENT, SHALL THE COMPANY OR ITS EMPLOYEES, AGENTS, LICENSORS OR CONTRACTORS BE LIABLE FOR ANY INCIDENTAL, INDIRECT, SPECIAL OR CONSEQUENTIAL DAMAGES ARISING OUT OF OR IN CONNECTION WITH THIS LICENSE OR THE SOFTWARE, INCLUDING, WITHOUT LIMITATION, LOSS OF USE, LOSS OF DATA, LOSS OF INCOME OR PROFIT, OR OTHER LOSSES SUSTAINED AS A RESULT OF INJURY TO ANY PERSON, OR LOSS OF OR DAMAGE TO PROPERTY, OR CLAIMS OF THIRD PARTIES, EVEN IF THE COMPANY OR AN AUTHORIZED REPRESENTATIVE OF THE COMPANY HAS BEEN ADVISED OF THE POSSIBILITY OF SUCH DAMAGES.** SOME JURISDICTIONS DO NOT ALLOW THE LIMITATION OF DAMAGES IN CERTAIN CIRCUMSTANCES, SO THE ABOVE LIMITATIONS MAY NOT ALWAYS APPLY.

6. **GENERAL:** THIS AGREEMENT SHALL BE CONSTRUED IN ACCORDANCE WITH THE LAWS OF THE UNITED STATES OF AMERICA AND THE STATE OF NEW YORK, APPLICABLE TO CONTRACTS MADE IN NEW YORK, AND SHALL BENEFIT THE COMPANY, ITS AFFILIATES AND ASSIGNEES. This Agreement is the complete and exclusive statement of the agreement between you and the Company and supersedes all proposals, prior agreements, oral or written, and any other communications between you and the company or any of its representatives relating to the subject matter. If you are a U.S. Government user, this Software is licensed with "restricted rights" as set forth in subparagraphs (a)–(d) of the Commercial Computer-Restricted Rights clause at FAR 52.227–19 or in subparagraphs (c)(1)(ii) of the Rights in Technical Data and Computer Software clause at DFARS 252.227–7013, and similar clauses, as applicable.

7. **ACKNOWLEDGEMENT:** YOU ACKNOWLEDGE THAT YOU HAVE READ THIS AGREEMENT, UNDERSTAND IT, AND AGREE TO BE BUND BY ITS TERMS AND CONDITIONS. YOU ALSO AGREE THAT THIS AGREEMENT IS THE COMPLETE AND EXCLUSIVE STATEMENT OF THE AGREEMENT BETWEEN YOU AND THE COMPANY.